Undergraduate Texts in Mathematics

Undergraduate Texts in Mathematics

Anglin: Mathematics: A Concise History and Philosophy.
Readings in Mathematics.
Apostol: Introduction to Analytic Number Theory. Second edition.
Armstrong: Groups and Symmetry.
Armstrong: Basic Topology.
Bak/Newman: Complex Analysis.
Banchoff/Wermer: Linear Algebra Through Geometry. Second edition.
Berberian: A First Course in Real Analysis.
Brémaud: An Introduction to Probabilistic Modeling.
Bressoud: Factorization and Primality Testing.
Bressoud: Second Year Calculus.
Readings in Mathematics.
Brickman: Mathematical Introduction to Linear Programming and Game Theory.
Cederberg: A Course in Modern Geometries.
Childs: A Concrete Introduction to Higher Algebra.
Chung: Elementary Probability Theory with Stochastic Processes. Third edition.
Cox/Little/O'Shea: Ideals, Varieties, and Algorithms.
Croom: Basic Concepts of Algebraic Topology.
Curtis: Linear Algebra: An Introductory Approach. Fourth edition.
Devlin: The Joy of Sets: Fundamentals of Contemporary Set Theory. Second edition.
Dixmier: General Topology.
Driver: Why Math?
Ebbinghaus/Flum/Thomas: Mathematical Logic. 2nd Edition.
Edgar: Measure, Topology, and Fractal Geometry.
Fischer: Intermediate Real Analysis.
Flanigan/Kazdan: Calculus Two: Linear and Nonlinear Functions. Second edition.
Fleming: Functions of Several Variables. Second edition.
Foulds: Optimization Techniques: An Introduction.
Foulds: Combinatorial Optimization for Undergraduates.
Franklin: Methods of Mathematical Economics.
Halmos: Finite-Dimensional Vector Spaces. Second edition.
Halmos: Naive Set Theory.
Hämmerlin/Hoffmann: Numerical Mathematics.
Readings in Mathematics.
Iooss/Joseph: Elementary Stability and Bifurcation Theory. Second edition.
Isaac: The Pleasures of Probability.
Readings in Mathematics.
James: Topological and Uniform Spaces.
Jänich: Linear Algebra.
Jänich: Topology.
Kemeny/Snell: Finite Markov Chains.
Klambauer: Aspects of Calculus.
Kinsey: Topology of Surfaces.
Lang: A First Course in Calculus. Fifth edition.

(continued after index)

Jack Macki Aaron Strauss

Introduction to Optimal Control Theory

With 70 Figures

Springer-Verlag
New York Berlin Heidelberg London Paris
Tokyo Hong Kong Barcelona Budapest

Jack Macki
Department of Mathematics
The University of Alberta
Edmonton, Canada T6G 2G1

Aaron Strauss
formerly of
Division of Mathematical and
Physical Sciences and Engineering
University of Maryland
College Park, MD 20742
USA

AMS Subject Classification: 49-01

Library of Congress Cataloging in Publication Data
Macki, Jack.
Introduction to optimal control theory.
(Undergraduate texts in mathematics)
Bibliography: p.
Includes index.
1. Control theory. I. Strauss, Aaron.
II. Title. III. Series.
QA402.3.M317 629.8′312 81-8889
AACR2

Printed on acid-free paper.

Printed and bound by Braun-Brumfield, Ann Arbor, MI.
Printed in the United States of America.

9 8 7 6 5 4 3 2 (Corrected second printing, 1995)

ISBN 0-387-90624-X Springer-Verlag New York Berlin Heidelberg
ISBN 3-540-90624-X Springer-Verlag Berlin Heidelberg New York

Dedicated to the memory of Aaron Strauss
(1940–1977)

Preface

This monograph is an introduction to optimal control theory for systems governed by vector ordinary differential equations. It is not intended as a state-of-the-art handbook for researchers. We have tried to keep two types of reader in mind: (1) mathematicians, graduate students, and advanced undergraduates in mathematics who want a concise introduction to a field which contains nontrivial interesting applications of mathematics (for example, weak convergence, convexity, and the theory of ordinary differential equations); (2) economists, applied scientists, and engineers who want to understand some of the mathematical foundations of optimal control theory.

In general, we have emphasized motivation and explanation, avoiding the "definition-axiom-theorem-proof" approach. We make use of a large number of examples, especially one simple canonical example which we carry through the entire book. In proving theorems, we often just prove the simplest case, then state the more general results which can be proved. Many of the more difficult topics are discussed in the "Notes" sections at the end of chapters and several major proofs are in the Appendices. We feel that a solid understanding of basic facts is best attained by at first avoiding excessive generality.

We have not tried to give an exhaustive list of references, preferring to refer the reader to existing books or papers with extensive bibliographies. References are given by author's name and the year of publication, e.g., Waltman [1974].

Prerequisites for reading this monograph are basic courses in ordinary differential equations, linear algebra, and modern advanced calculus (including some Lebesgue integration). Some functional analysis is used, but the proofs involved may be treated as optional. We have summarized the relevant facts from these areas in an Appendix. We also give references in this Appendix to standard texts in these areas.

We would like to express our appreciation to: Professor Jim Yorke of the University of Maryland for providing several important and original proofs to simplify the presentation of difficult material; Dr. Stephen Lewis of the University of Alberta for providing several interesting examples from Economics; Ms. Peggy Gendron of the University of Minnesota and June Talpash and Laura Thompson of Edmonton, Alberta for their excellent typing work; the universities (and, ultimately, the relevant taxpayers) of Alberta, Maryland, and Minnesota – the first two for their direct financial support, and the last for providing facilities for J.W.M. while on sabbatical; The National Research Council of Canada, for its continuing support of J.W.M.

Edmonton, Alberta
August, 1980

Jack W. Macki

Acknowledgement for the Second Printing

I would like to thank the following people for suggesting improvements and noting errors and oversights in the first printing: Yosef Cohen, Robert Elliott, Philip Loewen, Sim McKinney, Ruth Curtain and Hans Niewenhuis.

Contents

List of Symbols xi

Chapter I
Introduction and Motivation 1

1 Basic Concepts 1
2 Mathematical Formulation of the Control Problem 4
3 Controllability 6
4 Optimal Control 9
5 The Rocket Car 10
Exercises 15
Notes 20

Chapter II
Controllability 24

1 Introduction: Some Simple General Results 24
2 The Linear Case 28
3 Controllability for Nonlinear Autonomous Systems 38
4 Special Controls 44
Exercises 48
Appendix: Proof of the Bang–Bang Principle 50

Chapter III
Linear Autonomous Time-Optimal Control Problems 57

1 Introduction: Summary of Results 57
2 The Existence of a Time-Optimal Control; Extremal Controls; the Bang–Bang Principle 60

3 Normality and the Uniqueness of the Optimal Control 65
4 Applications 74
5 The Converse of the Maximum Principle 77
6 Extensions to More General Problems 79
Exercises 80

Chapter IV
Existence Theorems for Optimal Control Problems 82

1 Introduction 82
2 Three Discouraging Examples. An Outline of the Basic Approach to Existence Proofs 83
3 Existence for Special Control Classes 88
4 Existence Theorems under Convexity Assumptions 91
5 Existence for Systems Linear in the State 97
6 Applications 98
Exercises 100
Notes 102

Chapter V
Necessary Conditions for Optimal Controls—The Pontryagin Maximum Principle 103

1 Introduction 103
2 The Pontryagin Maximum Principle for Autonomous Systems 104
3 Applying the Maximum Principle 111
4 A Dynamic Programming Approach to the Proof of the Maximum Principle 118
5 The PMP for More Complicated Problems 124
Exercises 128

Appendix to Chapter V—A Proof of the Pontryagin Maximum Principle 134
Mathematical Appendix 147
Bibliography 160
Index 163

List of Symbols

$\mathscr{B}(\mathbf{x};\alpha)$	the open ball about $\mathbf{x}$ of radius α, $\{\mathbf{y} \mid \lvert\mathbf{y}-\mathbf{x}\rvert<\alpha\}$
C, $C[\mathbf{u}(\cdot)]$	cost function
$C(\mathbf{x}_0)$	cost function evaluated along the optimal control and response from $\mathbf{x}_0$
C^1	class of functions having continuous first partial derivatives
C^∞	functions having continuous partials of every order
C_0^∞	functions from C^∞ which have compact support
$\mathrm{co}(\Omega)$	convex hull of Ω
$\mathscr{C}$	controllable set
$\mathscr{C}(t)$	controllable set at time t
$\mathscr{C}_{\mathrm{BB}}$	controllable set using bang–bang controls
$\mathscr{C}_{\mathrm{BB}}(t)$	controllable set at time t using bang–bang controls
$\mathscr{C}_{\mathrm{BBPC}}$	controllable set using bang–bang piecewise constant controls
$d(\mathbf{x}, P)$	$\inf\{\lvert\mathbf{x}-\mathbf{y}\rvert : \mathbf{y}\in P\}$
∂S	boundary of the set S
f^0	$C[u(\cdot)]=\int_0^t f^0(s,\mathbf{x}[s],\mathbf{u}(s))\,ds$
$\mathbf{f}_\mathbf{x}$	matrix of partials $\partial f^i/\partial x^j$
$\hat{\mathbf{f}}$	$(f^0,\mathbf{f}^T)^T$; the extended velocity vector
$\mathrm{grad}_\mathbf{x} H$	gradient $=\left(\dfrac{\partial H}{\partial x^1},\dots,\dfrac{\partial H}{\partial x^n}\right)$

$h(P, Q)$	Hausdorff metric $= \inf\{\varepsilon : P \subset N(Q, \varepsilon)$ and $Q \subset N(P, \varepsilon)\}$
$H, H(\hat{\mathbf{w}}, \mathbf{x}, \mathbf{u})$	$\langle \hat{\mathbf{w}}, \hat{\mathbf{f}}(\mathbf{x}, \mathbf{u})\rangle$
Int S	the interior of the set S
$k(\tau)$	$\left\{\sum_{i=1}^{P} c^i[\hat{\mathbf{f}}(x_*[\tau], \mathbf{v}_i) - \hat{\mathbf{f}}(x_*[\tau], u_*(\tau))]c^i \geq 0, v_i \in \Psi, P \in N\right\}$; here $(x_*[\cdot], y_*(\cdot))$ is optimal and Ψ is the range set for admissible controls
$K(t; \mathbf{x}_0)$	reachable set at time t
$K_{\mathbf{BB}}(t; \mathbf{x}_0)$	reachable set at time t using bang-bang controls
$\mathcal{K}(t)$	$\left\{\sum_{i=1}^{P} c^i Y(t, \tau_i)\, \hat{\mathbf{z}}_i \mid c_i \geq 0, \hat{z}_i \in k(\tau_i)\right\}$, where $Y(t, \tau_i)$ is the fundamental matrix for (Lin) satisfying $Y(\tau_i, \tau_i) = I$
$\bar{\mathcal{K}}(t_1)$	$\{\hat{\mathbf{a}} + \beta\hat{\mathbf{b}} \mid \|\beta\| \leq \beta_0, \hat{\mathbf{a}} \in \mathcal{K}(t_1), \hat{\mathbf{b}} = \hat{\mathbf{f}}(\mathbf{x}_*[t_1], \mathbf{u}_*(t_1))\}$
(L)	$\dot{\mathbf{x}} = A(t)\mathbf{x} + B(t)\mathbf{u} + \mathbf{c}(t)$
(LA)	$\dot{\mathbf{x}} = A\mathbf{x} + B\mathbf{u}$
m	dimension of control vectors $\mathbf{u}(t)$
$M(\hat{\mathbf{w}}, \mathbf{x})$	$\sup\{H(\hat{\mathbf{w}}, \mathbf{x}, \mathbf{v}): \mathbf{v} \in \Omega\}$
$M \equiv (B, AB, \ldots, A^{n-1}B)$	the controllability matrix for (LA)
$N(P, \varepsilon)$	$\{\mathbf{x}: d(\mathbf{x}, P) < \varepsilon\}$
$o(\|\mathbf{x}\|)$	stands for any function $h(\mathbf{x})$ such that $\lim_{\mathbf{x}\to 0} \frac{h(\mathbf{x})}{\|\mathbf{x}\|} = 0$
$Q^+(t, \mathbf{x})$	$\{(y^0, \mathbf{y}) \mid \exists \mathbf{v} \in \Omega, \mathbf{y} = \mathbf{f}(t, \mathbf{x}, \mathbf{v}), y_0 \geq f^0(t, \mathbf{x}, \mathbf{v})\}$
RC	reachable cone, $\bigcup_{t>t_0} (t, K(t, \mathbf{x}_0))$
R^n	Euclidean n-dimensional space
sgn α	$\alpha/\|\alpha\|$ provided $\alpha \neq 0$
$\mathcal{T}(t)$	target state
$\mathcal{U}_{\mathbf{BB}}$	class of functions in $\mathcal{U}_m$ for which $\|u^i(t)\| = 1$
$\mathcal{U}_m$	$\bigcup_{t_1>t_0} \mathcal{U}_m(t_0, t_1)$
$\mathcal{U}_m(t_0, t_1)$	class of measurable functions from $[t_0, t_1]$ to Ω
$\mathcal{U}_{\mathbf{PC}}$	class of piecewise constant functions in $\mathcal{U}_m$
$\mathcal{U}_{\mathbf{PS}}$	class of piecewise smooth functions in $\mathcal{U}_m$
$\mathcal{U}_r$	class of piecewise constant functions in $\mathcal{U}_m$ with at most r discontinuities

$\mathcal{U}_\lambda$	class of functions in $\mathcal{U}_m$ having Lipschitz constant λ
$\mathcal{V}_m(t_0, t_1)$	class of measurable functions from $[t_0, t_1]$ to a given bounded set $\Psi \subset R^m$
$\mathcal{V}_m$	$\bigcup_{t_1 > t_0} \mathcal{V}_m(t_0, t_1)$
$\mathbf{x}(t; t_0, \mathbf{x}_0, \mathbf{u}(\cdot))$	solution of relevant differential equation through $\mathbf{x}_0$ at time t_0 corresponding to $\mathbf{u}(\cdot)$; the *state* vector
x^i	i^{th} component of $\mathbf{x}$
$\mathbf{x}^T$	transpose of $\mathbf{x}$
$\hat{\mathbf{x}}$	$(x^0, \mathbf{x}) \in R^{n+1}$; the extended state vector
$\langle \mathbf{x}, \mathbf{y} \rangle$	$\sum_i x^i y^i$
Δ	class of successful controls: they steer the initial state to the target
χ_Q	characteristic function of a set Q, i.e., $\chi = +1$ on Q, 0 on the complement of Q
Ω	the unit cube in R^m

Chapter I

Introduction and Motivation

1. Basic Concepts

In control theory, one is interested in governing the *state* of a *system* by using *controls*. The best way to understand these three concepts is through examples.

EXAMPLE I (A National Economy). The economy of a typical capitalistic nation is a *system* made up in part of the population (as consumers and as producers), companies, material goods, production facilities, cash and credit available, and so on. The *state* of the system can be thought of as a massive collection of data: wages and salaries, profits, losses, sales of goods and services, investment, unemployment, welfare costs, the inflation rate, gold and currency holdings, and foreign trade. The federal government can influence the state of this system by using several *controls*, notably the prime interest rate, taxation policy, and persuasion regarding wage and price settlements.

EXAMPLE II (Water Storage and Supply). As early as the third century B.C., systems similar to that sketched in Figure 1 were being used in water storage tanks. As the water level rises, the float will restrict the inlet flow; all inlet flow will cease when the water reaches a certain height. If water is withdrawn from the outlet at a certain rate then the float will tend to adjust the inlet flow so as to maintain the water height in the tank. One can think of the water in the tank along with the float, inlet, and outlet, as a *system*. The *control* is the position of the float. The *state* at any instant is a vector, consisting of the height of the water in the tank, the inlet rate of flow and the outlet rate of flow. In this example, the state of the system

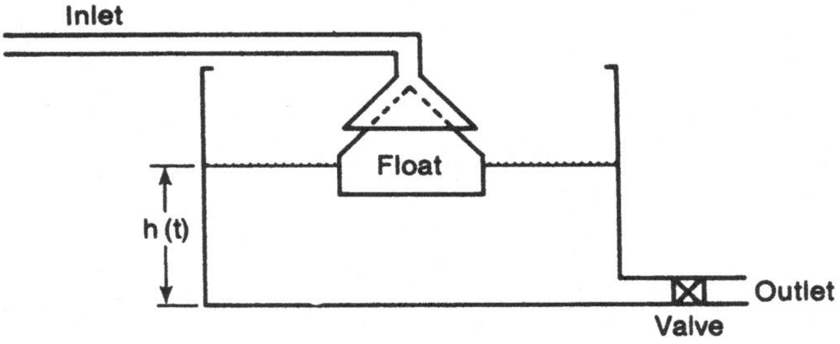

Figure 1

(rather than an external observer) automatically sets the control (position of the float) – this is an example of a *feedback* control system – the state is "fed back" to the control mechanism, which adjusts without outside influence.

We have chosen one example from economics and one from civil engineering. We could just as well have chosen examples from biology, economics, space flight, or several other fields, because the concepts of system, state, and control are so general. In the exercises at the end of the chapter we have given several more examples.

The essence of these examples is that in control theory we have a system and we try to influence the state of the system through controls. The *dynamics* of the system, that is, the manner in which the state changes under the influence of the controls, can be very complicated in real-world examples. In the case of a national economy, the dynamics is still a matter of considerable research. Of course, there are many general principles for a national economy – for example, raising the prime rate (a control) generally increases unemployment – but a detailed, accurate picture of the dynamics of a national economy is very difficult. On the other hand, the dynamics of the water storage system is relatively easy to describe. We won't do it here, since we are going to deal with an even simpler example shortly.

There are two remaining concepts to be described, namely the *constraints* on our controls, and the *objective* or *target* state(s) for our system. For a national economy, there are several obvious constraints on our controls, for example, taxation cannot be too excessive and the prime rate cannot be negative. There are also objective or target states – ideally a government wants a state of the economy with full employment, an inflation rate of 0%, low interest rates, and low taxes. In fact, they may have to settle for a realistic target state with an unemployment rate less than 8%, inflation less than 10%, moderate interest rates, and realistic tax rates. *Any* state with these properties would do, so there are many target states. In fact, the set of target states might vary with time, reflecting political and social changes.

In the water tank example, the constraint on the position of the stopper is that it always floats at a fixed distance above the water level; also the velocity with which the stopper can move is tied directly to the rate of change of the water level height. The objective might be a pre-set water height.

EXAMPLE 1 (The Rocket Car). This example will be used throughout this monograph to motivate and illustrate concepts and results. The car runs on rails on the level, has a mass of one, and is equipped with two rocket engines, one on each end (Figure 2). The problem is to move the car from

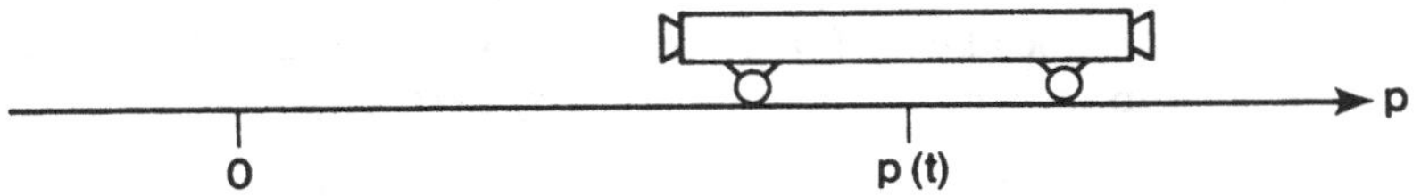

Figure 2

any given location to a fixed pre-assigned destination. For simplicity, we place the destination at the origin and denote the position of the center of the car by $p(t)$. If the car is at a position p_0 at time $t=0$, with velocity v_0, we want to fire the two engines according to some recipe (pattern, program) which will have us arrive at $p=0$ *at rest* (with velocity zero) at some instant $t_1>0$. We can take as our *system* the car plus its track; as the *state* we take the two-vector $\mathbf{x}(t)=(p(t), \dot{p}(t))$; the initial state (p_0, v_0) is assumed given. The physical reason for using a two-vector for the state is simple – we want to know where we are and how fast we are going. Our *target* state is (0, 0). A *control* $u(t)$ is a real-valued function, representing the force on the car due to firing either engine at time t. If we fire the right engine at time t^*, we will say the force is negative, if we fire the left engine we take the force positive.

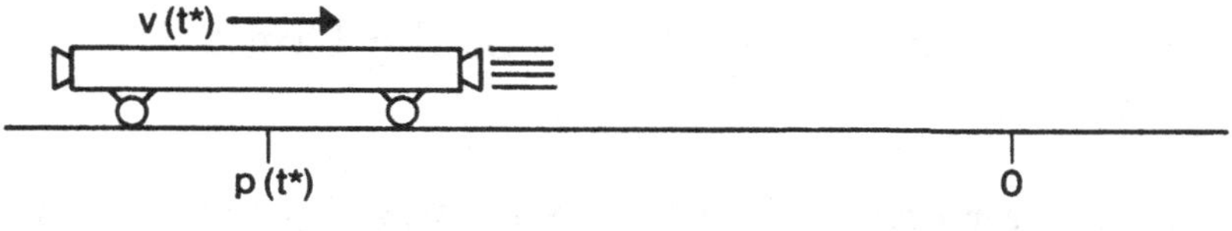

Figure 3 (Moving to the Right) The *Force* Is to the Left When $u(t^*)<0$.

Then the *dynamics* of our system is given by Newton's law $F=ma$, which can be written as $\ddot{p}(t)=u(t)$. This has the natural vector form

$$\mathbf{x}(t)=\begin{bmatrix} p(t) \\ \dot{p}(t) \end{bmatrix}, \qquad \dot{\mathbf{x}}(t)=\begin{bmatrix} 0 & 1 \\ 0 & 0 \end{bmatrix}\mathbf{x}(t)+u(t)\begin{bmatrix} 0 \\ 1 \end{bmatrix}.$$

There are *constraints* on the magnitude of $u(t)$, based on the size of the rocket motors and the amount of acceleration stress allowed on the car.

A mathematically reasonable assumption is that $u(t)$ is measurable and bounded, and we take our constraint to be $|u(t)|\leq 1$ for simplicity. Since measurable functions can be quite pathological, we will often use classes that are physically more reasonable, e.g., piecewise constant controls.

A given control function $u(t)$ is a recipe for firing our engines. For example

$$u(t)=\begin{cases}+1, & 0\leq t\leq 1;\\ -\frac{1}{2}, & 1<t\leq 3;\end{cases}$$

tells us to fire the left engine at full force for one unit of time, then fire the right engine at half force for two units of time.

If (p_0, v_0) is the position of the car at $t=0$, we can integrate the differential equation twice and then integrate by parts to get:

$$p(t)=p_0+v_0t+\int_0^t (t-r)u(r)\,dr, \qquad \dot{p}(t)=v_0+\int_0^t u(r)\,dr.$$

Thus each choice of control, $u(\cdot)$, generates a *response* $\mathbf{x}[t]\equiv\mathbf{x}(t;\mathbf{x}_0, u(\cdot))$. (We delete reference to the initial time t_0, since we always take it to be zero for simplicity. For systems not explicitly containing t, this is in fact not a restriction.) We use $u(\cdot)$ to refer to the function $u(t)$, on its domain of definition, as an entity. If the response $\mathbf{x}(t;\mathbf{x}_0, u(\cdot))$ reaches the target $(0, 0)$ at some $t_1>0$, then $u(\cdot)$ is a *successful control.* There might be no such control or many. When there are several successful controls, the choice of one over the other may be dictated by practicality, and/or by a *cost* or *performance criterion.* For example, later on we shall consider the criteria: (1) least time, (2) least energy expended, (3) least fuel expended. Our control problem will then become an *Optimal Control Problem.*

2. Mathematical Formulation of the Control Problem

We now give a precise mathematical formulation of the type of control problem we will be discussing. Let m, n be natural numbers, and let R stand for the real numbers. If $\mathbf{x}$, $\mathbf{y}$ are column vectors in R^n, we denote their i^{th} components by x^i, y^i respectively. We define $\mathbf{x}^T$ to be the transpose of $\mathbf{x}$, and introduce a dot product and two norms:

$$\langle\mathbf{x}, \mathbf{y}\rangle=\mathbf{x}^T\mathbf{y}=\sum_{i=1}^n x^iy^i,$$

$$|\mathbf{x}|=\sum_{i=1}^n |x^i|, \qquad \|\mathbf{x}\|=\langle\mathbf{x}, \mathbf{x}\rangle^{1/2}.$$

If we need to square a scalar-valued function $\phi(t)$, we will write $[\phi(t)]^2$, while $x^2(t)$ will denote the second component of the vector-valued function $\mathbf{x}(t)$ – in context the distinction will always be obvious. Let Ω denote the unit cube in R^m, i.e.,

$$\Omega = \{\mathbf{c} \mid \mathbf{c} \in R^m, |c^i| \leq 1, i = 1, 2, \ldots, m\}.$$

For $t_1 \geq 0$, define

$$\mathcal{U}_m[0, t_1] = \{\mathbf{u}(\cdot) \mid \mathbf{u}(t) \in \Omega \text{ and } \mathbf{u}(\cdot) \text{ measurable on } [0, t_1]\},$$

$\mathcal{U}_m = \bigcup_{t_1>0} \mathcal{U}_m[0, t_1]$. Unless explicitly stated otherwise, our controls $\mathbf{u}(\cdot)$ will always be assumed to belong to $\mathcal{U}_m$. This mildly cumbersome definition of our admissible controls allows each control $\mathbf{u}(\cdot)$ to have its own corresponding interval of definition $[0, t_1(\mathbf{u})]$.

We assume that for each $t \geq 0$ we are given a target set $\mathcal{T}(t) \subset R^n$ where $\mathcal{T}(t)$ is a closed set. For most of this monograph we will take $\mathcal{T}(t) \equiv 0 \in R^n$ for simplicity. Nevertheless, general target sets are important, as we mentioned in the example of a national economy.

We assume that the dynamics of the system, that is, the evolution of the state $\mathbf{x}(t)$ under a given control $\mathbf{u}(t)$, is determined by a vector ordinary differential equation:

$$(1) \qquad \dot{\mathbf{x}}(t) = \mathbf{f}(t, \mathbf{x}(t), \mathbf{u}(t)), \qquad \mathbf{x}(t_0) = \mathbf{x}_0.$$

We will always assume that $\mathbf{f}(t, \mathbf{x}, \mathbf{u})$, $\partial f^i/\partial x^j$, $\partial f^i/\partial u^k$ are all continuous $(i, j = 1, \ldots, n; k = 1, \ldots, m)$ on $[0, \infty) \times R^n \times R^m$, although most results are valid under weaker conditions. This assumption guarantees local existence and uniqueness of the solution of (1) for a given $\mathbf{u}(\cdot) \in \mathcal{U}_m$. Because $\mathbf{u}(\cdot)$ is only assumed measurable and bounded, the right side of the equation $\dot{\mathbf{x}} = \mathbf{f}(t, \mathbf{x}, \mathbf{u}(t))$ is continuous in $\mathbf{x}$ but only measurable and bounded in t for each $\mathbf{x}$. Therefore, solutions are understood to be absolutely continuous functions that satisfy (1) almost everywhere. The solution of (1) for a given $\mathbf{u}(\cdot)$ will be called the *response* to $\mathbf{u}(\cdot)$; we denote it by $\mathbf{x}[t] \equiv \mathbf{x}(t; \mathbf{x}_0, \mathbf{u}(\cdot))$. The *control problem* is to determine those $\mathbf{x}_0$ and $\mathbf{u}(\cdot) \in \mathcal{U}_m$ such that the associated response satisfies $\mathbf{x}[t_1] \in \mathcal{T}(t_1)$ for some $t_1 > 0$; we then say that *the control* $\mathbf{u}(\cdot)$ *steers* $\mathbf{x}_0$ *to the target.*

If the control $\mathbf{u}(\cdot)$ is defined on $[0, t_1)$ $(t_1 \leq +\infty)$, it is not assumed that the corresponding response extends to $[0, t_1)$; a given response $\mathbf{x}(t; \mathbf{x}_0, \mathbf{u}(\cdot))$ may only exist on some subinterval of $[0, t_1)$. For example, consider the scalar problem $\dot{x} = x^2 + u$, $x_0 = 1$. For $u_0(t) \equiv 0$ $(t_1 = +\infty)$, the response is $x(t; 1, u_0(\cdot)) = 1/(1-t)$, which only exists on $[0, 1)$. For linear equations,

$$(2) \qquad \dot{\mathbf{x}}(t) = A(t)\mathbf{x}(t) + B(t)\mathbf{u}(t),$$

with $A(t)$ and $B(t)$ continuous on $[0, t_1)$, solutions always extend to $[0, t_1)$.

Thus, our general control problem consists of a class of admissible *controls* $\mathcal{U}_m$, a vector differential equation (1) describing the *dynamics* of our system, and a family of *target* sets $\mathcal{T}(t)$. One basic problem is to describe those initial states $\mathbf{x}_0 \in R^n$ which can be steered to the target, that is, those states which are *controllable*. Also, if $\mathcal{T}(t)$ is a closed, bounded set with non-empty interior, we want the response to penetrate $\mathcal{T}(t)$ rather than graze $\mathcal{T}(t)$ tangentially, otherwise perturbations that are always present in the real world might pull the response away from the target. This problem comes under the heading of *transversality* conditions: we want the trajectory to be non-tangential (transverse) to the boundary of $\mathcal{T}(t)$.

If we know that a given $\mathbf{x}_0$ can be steered to the target, another problem is one of *synthesis* – to describe at least one control which will do the job; more generally, to describe a method for constructing a successful control for each initial state which can be steered to the target.

The term "synthesis" is most often used in the context of *feedback* controls. These are controls which do not depend explicitly on t, but instead depend only on the state of the system, as mentioned in the water storage example.

There is also the problem of a *practical* control. With the rocket car, for example, it is conceivable that one successful control might be an infinite sequence of alternate bursts on the two engines, the alternate bursts causing us to overshoot the origin by less and less. This of course is not a very practical way to get to the origin with zero velocity. Therefore, we will need to consider special classes of controls.

We will describe the mathematical formulation of *controllability* in the next section (Section 3). In Section 4, we will formulate the general *optimal control problem*. In Section 5, we will analyze these concepts in detail for the special case of the rocket car. In this last section, we will also deal with the problem of *synthesis* for the rocket car.

3. Controllability

Given a control problem:

$$(3) \qquad \dot{\mathbf{x}} = \mathbf{f}(t, \mathbf{x}, \mathbf{u}), \qquad \mathbf{u}(\cdot) \in \mathcal{U}_m, \qquad \mathcal{T}(t) \text{ given,}$$

we want to discuss the problem of controllability, that is, we wish *to describe those initial states* $\mathbf{x}(0) = \mathbf{x}_0$ *for which at least one successful control* $\mathbf{u}(\cdot)$ *exists*. We define the *controllable set* $\mathscr{C} = \bigcup_{t_1>0} \mathscr{C}(t_1)$, where

$$\mathscr{C}(t_1) = \{\mathbf{x}_0 \in R^n \mid \exists \mathbf{u}(\cdot) \in \mathcal{U}_m \text{ such that } \mathbf{x}(t_1; \mathbf{x}_0, \mathbf{u}(\cdot)) \in \mathcal{T}(t_1)\}$$

is simply those states which can be steered to the target at time t_1.

Two major problems in controllability theory can now be formulated:

(a) to describe $\mathscr{C}$;
(b) to describe how $\mathscr{C}$ changes if we vary the control set $\mathcal{U}_m$.

Consider a real-world process which has reached its target. The process may be continually subjected to small disturbances that randomly push it away from the target state. Therefore, real-world situations require that we be able to steer all nearby states to the target. It is thus essential that *$\mathscr{C}$ contain a full neighborhood of the target.* Of course, it would be even better if $\mathscr{C} = R^n$. We will consider these two questions in detail in Chapter II.

Concerning special control classes, there are three subsets of $\mathscr{U}_m$ that we will consider:

(a) $$\mathscr{U}_{\text{PC}}[0, t_1] = \{\mathbf{u}(\cdot) \in \mathscr{U}_m[0, t_1] \mid \mathbf{u}(\cdot) \text{ piecewise constant on } [0, t_1]\}$$

where by $\mathbf{u}(\cdot)$ piecewise constant we mean that there exists a partition (depending on $\mathbf{u}(\cdot)$) $0 = s_0 < s_1 < \cdots < s_l = t_1$ such that $\mathbf{u}(t)$ is constant on each interval $[s_{k-1}, s_k)$. From a practical point of view, $\mathscr{U}_{\text{PC}} = \bigcup_{t_1>0} \mathscr{U}_{\text{PC}}[0, t_1]$ contains controls which are much easier to use than a general control from $\mathscr{U}_m$.

(b) $$\mathscr{U}_\varepsilon[0, t_1] = \{\mathbf{u}(\cdot) \in \mathscr{U}_m[0, t_1] \mid \mathbf{u}(\cdot) \text{ absolutely continuous}, \mathbf{u}(0) = \mathbf{u}(t_1) = 0, \\ |\mathbf{u}(t)| \leq 1 \text{ and } |\dot{\mathbf{u}}(t)| \leq \varepsilon \text{ a.e. on } [0, t_1]\},$$

where $\varepsilon > 0$ is fixed. $\mathscr{U}_\varepsilon = \bigcup_{t_1>0} \mathscr{U}_\varepsilon[0, t_1]$ is the class of *smooth* controls which do not change abruptly – an example is the steering of a car.

(c) $$\mathscr{U}_{\text{BB}}[0, t_1] = \{\mathbf{u}(\cdot) \in \mathscr{U}_m[0, t_1] \mid |u^i(t)| \equiv 1 \text{ on } [0, t_1], i = 1, \ldots, m\},$$

is the class of "bang–bang" controls. Controls from the class $\mathscr{U}_{\text{BB}} = \bigcup_{t_1>0} \mathscr{U}_{\text{BB}}[0, t_1]$ use maximum allowable power (remember $\mathbf{u}(t) \in \Omega$) at all times. These controls might be easier to synthesize for our rocket car, for example, since our engines would only need two settings, "off" and "full power." We do not require that a function $\mathbf{u}(\cdot) \in \mathscr{U}_{\text{BB}}$ be piecewise constant, therefore $\mathscr{U}_{\text{BB}}$ contains some rather pathological functions. The class of piecewise constant bang–bang controls on $[0, t_1]$ is denoted $\mathscr{U}_{\text{BBPC}}[0, t_1]$ and $\mathscr{U}_{\text{BBPC}} = \bigcup_{t_1>0} \mathscr{U}_{\text{BBPC}}[0, t_1]$.

We will discuss these special control classes in detail in Chapter II; for now we will deal only with $\mathscr{U}_m$.

To study controllability for a given initial state $\mathbf{x}_0$, it is useful to define $K(t; \mathbf{x}_0)$, *the reachable set at time t*, and $\text{RC}(\mathbf{x}_0)$, the *reachable cone*. $K(t; \mathbf{x}_0)$ is the set of states in R^n which can be reached at time t, beginning from state $\mathbf{x}_0$ at time $t_0 = 0$, using all possible admissible controls:

$$K(t; \mathbf{x}_0) = \{\mathbf{x}(t; \mathbf{x}_0, \mathbf{u}(\cdot)) \mid \mathbf{u}(\cdot) \in \mathscr{U}_m\};$$

and

$$\text{RC}(\mathbf{x}_0) = \{(t, \mathbf{x}(t; \mathbf{x}_0, \mathbf{u}(\cdot))) \mid t \geq 0, \mathbf{u}(\cdot) \in \mathscr{U}_m\} = \bigcup_{t\geq 0} \{t\} \times K(t; \mathbf{x}_0).$$

For simplicity, we have taken $t_0 = 0$ (if we had allowed t_0 to be arbitrary, then we would have to define $K(t; t_0, \mathbf{x}_0)$, RC $(t_0; \mathbf{x}_0)$ in the obvious way). Figure 4 is a hypothetical sketch of $K(t; \mathbf{x}_0)$, RC $(\mathbf{x}_0)$ for $n = 2$. Two responses have been sketched in, giving an idea of the boundary of RC $(\mathbf{x}_0)$.

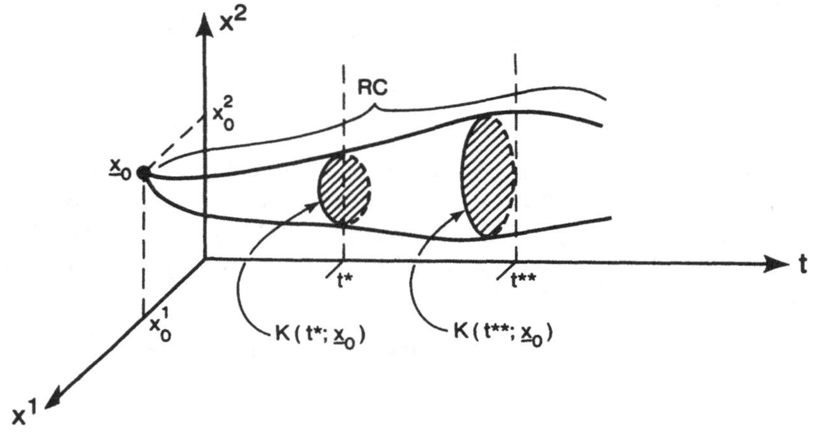

Figure 4

RC $(\mathbf{x}_0)$ looks like a cone and $K(t; \mathbf{x}_0)$ is a slice of the cone at time t. The sets $K(t; \mathbf{x}_0)$ are in fact defined in (x^1, x^2) space, so we should project them back onto the (x^1, x^2) plane, as sketched in Figure 5. We are then looking directly at the evolution of the reachable states as time passes (beginning always from $\mathbf{x}_0$ at $t_0 = 0$). In Chapter II we will study the sets $K(t; \mathbf{x}_0)$, RC $(\mathbf{x}_0)$ in detail; in Section 5 of this chapter we will describe $K(t; \mathbf{x}_0)$, RC $(\mathbf{x}_0)$ for the rocket car.

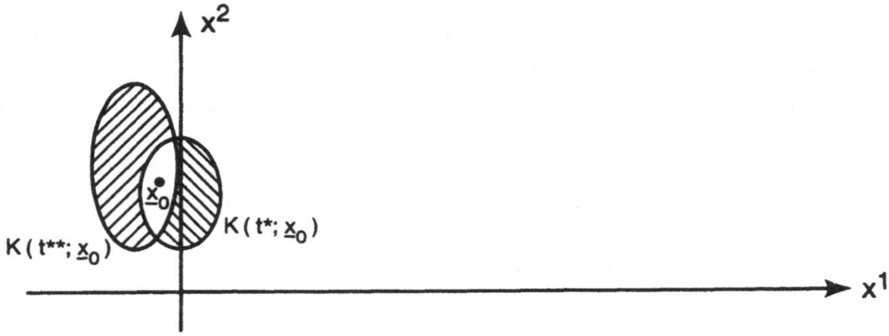

Figure 5

4. Optimal Control

The basic control problem may have associated with it a cost functional or performance criterion. We will only deal with cost functionals of the form

$$C[\mathbf{u}(\cdot)] = \int_0^{t_1} f^0(\mathbf{x}[t], \mathbf{u}(t))\, dt, \qquad \mathbf{x}[t] \equiv \mathbf{x}(t; \mathbf{x}_0, \mathbf{u}(\cdot)),$$

where f^0 is a given real-valued function. The *optimal control problem* is to steer $\mathbf{x}_0$ to a state in the target, using a control $\mathbf{u}(\cdot)$ from the appropriate class for the problem, in such a way that $C[\mathbf{u}(\cdot)]$ is a minimum. More precisely, let the successful controls be denoted by Δ, i.e.,

$$\Delta = \{\mathbf{u}(\cdot) \in \mathcal{U}_m \mid \exists t_1 \geq 0 \text{ such that } \mathbf{x}(t_1; \mathbf{x}_0, \mathbf{u}(\cdot)) \in \mathcal{T}(t_1)\}.$$

Then a control $\mathbf{u}_*(\cdot) \in \mathcal{U}_m$ is optimal if it is successful, i.e., $\mathbf{u}_*(\cdot) \in \Delta$, and

$$C(\mathbf{u}_*(\cdot)) \leq C(\mathbf{u}(\cdot)) \quad \text{for all } \mathbf{u}(\cdot) \in \Delta.$$

In optimal control theory, there are two basic problems:

(a) to prove the *existence* of an optimal control, and
(b) to *synthesize* an optimal control, that is, give a recipe for steering to the target in an optimal manner.

There are several other related problems, e.g., uniqueness of the optimal control and its practicality. Finally, there is the question of *necessary conditions* which an optimal control must satisfy. Here there is a strong analogy with the calculus of variations. In the calculus of variations, the real-valued function $y(\cdot)$ which minimizes the real-valued functional $\int_a^b g^0(t, y(t), \dot{y}(t))\, dt$, over (say) the class of continuously differentiable functions satisfying $y(a) = y(b) = 0$, must be a solution of the Euler–Lagrange equation

$$\frac{d}{dt}\left[\frac{\partial g^0}{\partial \dot{y}}(t, y(t), \dot{y}(t))\right] - \frac{\partial g^0}{\partial y}(t, y(t), \dot{y}(t)) = 0.$$

This turns out to be enormously useful in searching for the minimizing function, since the solutions of the Euler–Lagrange equation are often easy to describe. In optimal control theory, the analogous necessary condition for an optimal control $\mathbf{u}(\cdot)$ is known as the *Pontryagin Maximum Principle*. Chapter V is devoted to a description and proof of this principle. In the case of linear control problems, for example, the maximum principle will show that, under reasonable restrictions, the optimal control is bang–bang and piecewise constant. The connections between optimal control theory and the calculus of variations are further discussed in the Notes at the end of this chapter.

5. The Rocket Car

In this section we will work through the solution of the rocket car problem, to illustrate $K(t; \mathbf{x}_0)$, RC $(\mathbf{x}_0)$, and the concepts of controllability, synthesis, and optimality. All of the theoretical results derived in this monograph will eventually be applied to this example. In addition, many of these results will be motivated by first showing their validity for the case of the rocket car. For example, we will show below that the class of bang–bang controls $\mathscr{U}_{\mathrm{BB}}$ is as effective as the general class of controls $\mathscr{U}_m$ for the case of the rocket car, which (hopefully) motivates the general bang–bang principle to be presented in Chapter II.

We will use $(p(t), q(t))$, the position and velocity, as our state vector coordinates rather than (x^1, x^2), so the dynamics is described by

$$\text{(R)} \quad \dot{p}(t) = q(t), \qquad \dot{q}(t) = u(t), \qquad -1 \leq u(t) \leq 1, \qquad u(\cdot) \text{ measurable,}$$

with $(p(t), q(t))$ specified as some (p_0, q_0) at $t = 0$. Our target is $\mathscr{T}(t) \equiv (0, 0)$; we want $(p(t_1), q(t_1)) = (0, 0)$, where $t_1 > 0$ is not specified. In defining an appropriate cost functional, there are several considerations we wish to keep in mind:

(a) The entire process must take place within a reasonable (but not precisely specified) period of time. This "performance criterion" is measured by $\int_0^{t_1} dt$. (The upper limit t_1 is the time at which the particular response reaches the origin, and in general depends on $u(\cdot)$.)
(b) The kinetic energy allowed in the system should be limited in order to control wear and tear on the machinery (e.g., bearings). One way to measure this is to use $\int_0^{t_1} [q(t)]^2 \, dt$.
(c) The expenditure of fuel must be kept within reasonable (but not precisely specified) limits. This might be measured by $\int_0^{t_1} |u(t)| \, dt$ (assuming fuel expended is proportional to force developed).

To incorporate *all* of the above performance criteria, we define

$$C[u(\cdot)] = \int_{t_0}^{t_1} (\lambda_1 + \lambda_2 [q(t)]^2 + \lambda_3 |u(t)|) \, dt$$

where $\lambda_k \geq 0$ $(k = 1, 2, 3)$, $\lambda_1 + \lambda_2 + \lambda_3 = 1$. Each λ_k represents the proportional importance of the cost which it multiplies. The vector differential equation (R) is linear, so for each control $u(\cdot)$ and initial state (p_0, q_0) there is a unique response, $\mathbf{x}[t]$, defined for all $t \geq 0$. We first investigate the rocket car optimal control problem for $u(\cdot) \in \mathscr{U}_{\mathrm{BBPC}}$. The corresponding reachable set will be labelled $K_{\mathrm{BBPC}}(t; \mathbf{x}_0)$, to distinguish it from the reachable set $K(t; \mathbf{x}_0)$ when we are allowed to use any control from $\mathscr{U}_1$.

If $u(t) \equiv 1$ on some time interval starting at $t = 0$, then

$$(+) \qquad \dot{p} = q, \qquad \dot{q} = 1 \Rightarrow q\dot{q} = \dot{p} \Rightarrow [q(t)]^2 - q_0^2 = 2[p(t) - p_0];$$

if $u(t) \equiv -1$ on such an interval, then

$$(-) \qquad \dot{p} = q, \qquad \dot{q} = -1 \Rightarrow q\dot{q} = -\dot{p} \Rightarrow [q(t)]^2 - q_0^2 = -2[p(t) - p_0].$$

Therefore the corresponding responses move on the parabolas $\pm 2p = q^2 + \alpha$ where α is a constant (Figure 6). (To emphasize the parabolas, we have sketched the untraversed parts with dotted lines.)

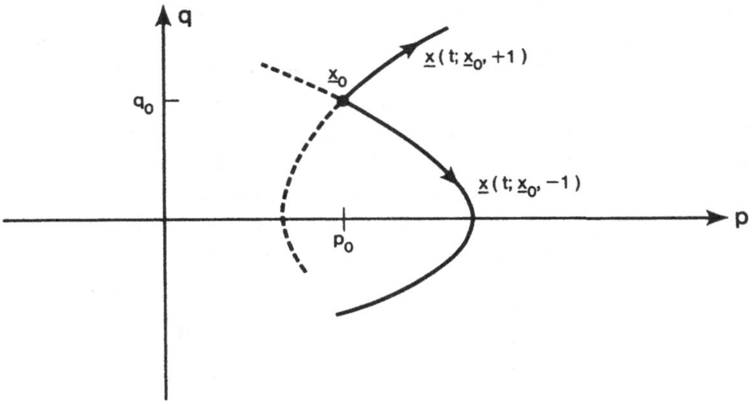

Figure 6

If we alternately set $u(t)$ equal to ± 1 on several successive intervals of time, we get trajectories (responses) as in Figures 7(a), (b). The sign of the corresponding control is in parentheses beside the trajectory. The points in (p, q)-space where we switch values of the control (*switching points*) are labelled by the symbol □. We have sketched in the untravelled parts of some parabolas with dots to help visualize the various parabolic trajectories. Notice that when $q = 0$, then $\dot{p} = 0$ and $\ddot{p} = \dot{q} = \pm 1$, so p will be a local min or max.

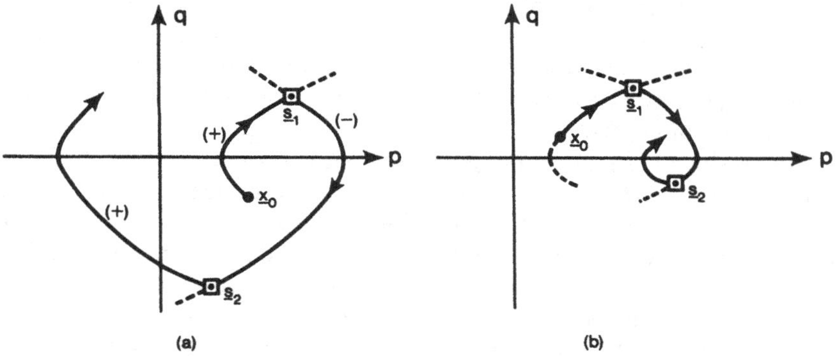

Figure 7

The differential equation (R) is autonomous (does not explicitly involve t). Therefore, the response of the system to $u(t) \equiv +1$ (say) does not depend on the particular time when we switch to this control from $u(t) \equiv -1$ (the switching time); the response depends only on the *state* when we switch. In Figure 7(a), the state at the point labelled $\mathbf{s}_1$ determines the subsequent path, and it does not matter whether the time when the system reaches $\mathbf{s}_1$ is 10^8 or 10^{-8}. We have labelled the same state with the symbol $\mathbf{s}_1$ in both Figures 7(a) and 7(b); the subsequent trajectories are identical.

To describe the reachable set $K_{\text{BBPC}}(t_1; \mathbf{x}_0)$, we fix $t_1 > 0$ and take $\mathbf{x}_0 = (p_0, 0)$ for simplicity. In Figure 8(a) we have sketched the two respective basic responses to $u(t) \equiv -1$ on $[0, t_1]$, and $u(t) \equiv +1$ on $[0, t_1]$. In Figure 8(b) we have sketched all those states which can be reached by switching once within a fixed time interval $[0, t_1]$. Of course, the farther you go along a basic response from Figure 8(a) before switching, the less time you have to follow the new trajectory after the switch, thus the football shape.

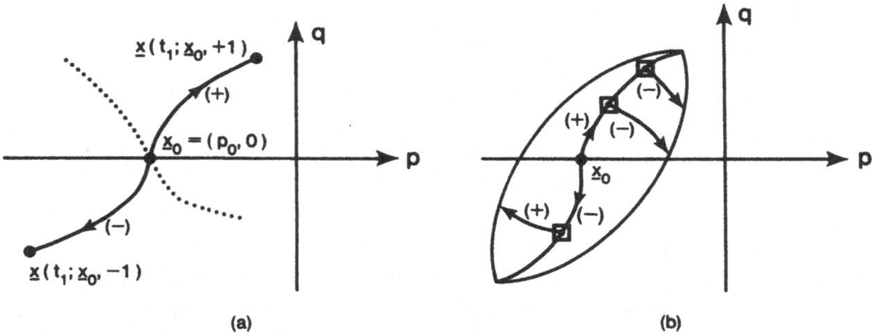

Figure 8

The boundary of the "football" in Figure 8(b) is the set of reachable points $\mathbf{x}(t_1; \mathbf{x}_0, u(\cdot))$, for one-switch controls. It turns out that $K_{\text{BBPC}}(t_1; \mathbf{x}_0)$ is *exactly* the entire closed region inside this boundary. We will show how to arrive at any interior point of the "football" at exactly time t_1, using a piecewise constant bang–bang control, thus $K_{\text{BBPC}}(t_1; \mathbf{x}_0)$ is at least the entire "football." We will not prove that responses to piecewise constant bang–bang controls always remain inside the "football," since at this stage it would be quite involved.

To convince the reader that given any point $\mathbf{z}$ inside the "football," we can arrange to arrive at $\mathbf{z}$ at exactly the instant $t = t_1$ using piecewise constant bang–bang controls, we shall show how to "waste" arbitrary amounts of time. Referring to Figure 9, given the desired state $\mathbf{z}$ we go backwards in time along a (−) parabola until we hit the main (+) parabola. Because the system is autonomous, it takes a certain *fixed* amount of time (less than t_1) to go from the state $\mathbf{x}_0 = (p_0, 0)$ "up" to the state □ and "down" to the state $\mathbf{z}$, so we need to waste some time. The idea is to initially run around

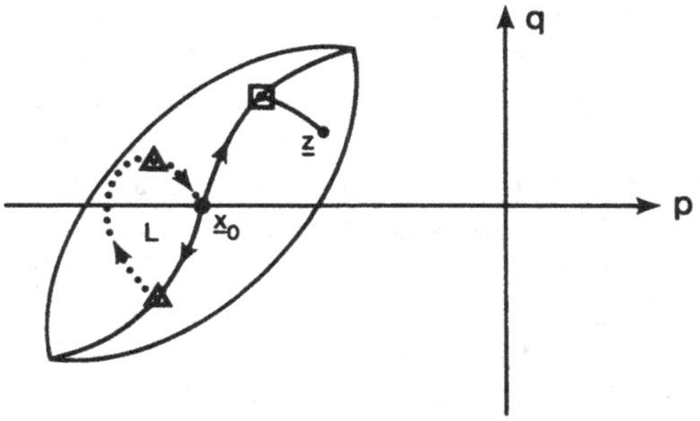

Figure 9

the loop $\mathscr{L}$ (in dots) until we have wasted just the right amount of time (ending up at $\mathbf{x}_0$), then proceed to the state $\square$ and $\mathbf{z}$. Since one can sketch in a continuum of loops ($\mathscr{L}$) made up of pieces of (+) and (−) parabolas, one can always find a loop such that a finite number of trips around it will "waste" a given length of time.

Surprisingly, the football shaped region of Figure 9 is also $K(t_1; \mathbf{x}_0)$, that is, we get exactly the same reachable set for *general* controls from $\mathscr{U}_1[0, t_1]$. This is in fact a specific instance of the bang–bang principle, to be discussed in Chapter II.

The description of $K_{\mathrm{BBPC}}(t_1; \mathbf{x}_0)$ for an initial state $\mathbf{x}_0 = (p_0, v_0)$ with $v_0 \neq 0$ is only slightly more complicated, and we omit it. We turn instead to a description of the controllable set $\mathscr{C}_{\mathrm{BBPC}}$ for the rocket car, i.e., those initial states which can be steered to the zero state by bang–bang piecewise constant controls.

$$\mathscr{C}_{\mathrm{BBPC}} = \{\mathbf{x}_0 | \exists \mathbf{u}(\cdot) \in \mathscr{U}_{\mathrm{BBPC}} \text{ and } t_1 > 0 \text{ such that } \mathbf{x}(t_1, \mathbf{x}_0, \mathbf{u}(\cdot)) = \mathbf{0}\}.$$

We will show that $\mathscr{C}_{\mathrm{BBPC}} = R^2$, and in fact we will need at most one switch. Referring to Figure 10, we first sketch the response $[q]^2 = 2p$ to $u(t) \equiv +1$

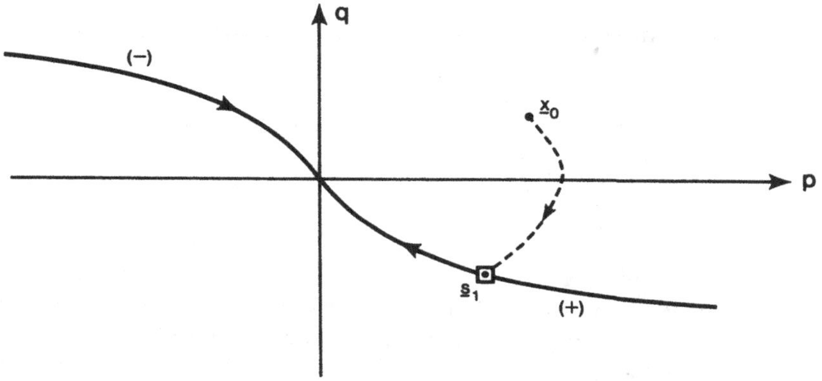

Figure 10 (+) (−) Is the Switching Curve

for $t \in [0, \infty)$ which passes through (0, 0); similarly we sketch the response $[q]^2 = -2p$ to $u(t) \equiv -1$ for $t \in [0, \infty)$. These two responses (labelled (+), (−), respectively, in Figure 10) of course pass through our target (0, 0), so given any state $\mathbf{x}_0 = (p_0, q_0)$, we simply try to steer to one of these two basic trajectories. This can always be done, as a little experimentation shows. Because the dynamics is governed by an autonomous differential equation, we can specify a synthesis that depends only on the *state* (a *feedback* control): if $\mathbf{x}_0$ is above the union of (+) and (−) from Figure 10, set $u(t) \equiv -1$ until the state reaches the curve (+) ($\mathbf{s}_1$ in the sketch) then switch to $u(t) \equiv +1$. If $\mathbf{x}_0$ is below the union of (+) and (−), set $u(t) \equiv +1$ until the state reaches the curve (−), then switch to $u(t) \equiv -1$. Notice that in all cases, $u(t)$ depends only on the state $\mathbf{x}(t)$.

We have therefore shown that $\mathscr{C}_{\mathrm{BBPC}} = R^2$, and we have a successful synthesis of a practical feedback control regime for the rocket car, which can be represented by a sketch in state-space (Figure 11). This need not be the only successful control scheme, however.

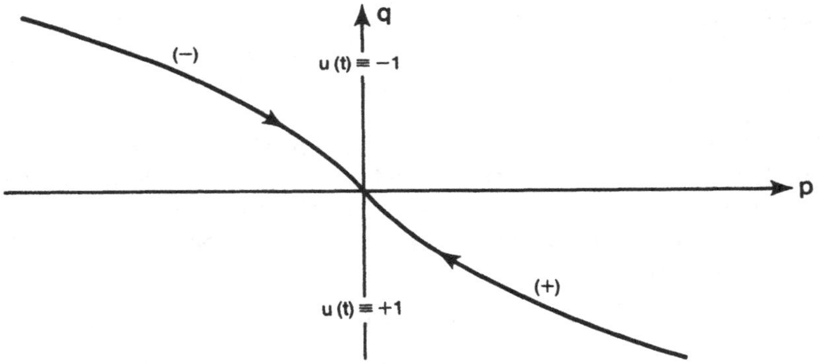

Figure 11 Synthesis Recipe for the Rocket Car

We will briefly discuss the *optimal* control problem for the rocket car. The cost functional is

$$C[u(\cdot)] = \int_0^{t_1} \{\lambda_1 + \lambda_2[q(t)]^2 + \lambda_3|u(t)|\}\, dt,$$

$$\sum_1^3 \lambda_k = 1, \qquad \lambda_k \geq 0 \quad \text{for } k = 1, 2, 3,$$

where we shall allow $u(\cdot) \in \mathscr{U}_1$. We know that a given initial state $\mathbf{x}_0$ can be steered to $\mathbf{0} \in R^2$ using $u(t) = \pm 1$ with one switch. However, it is conceivable that $C[u(\cdot)]$ is a minimum for some other successful control from $\mathscr{U}_1$ – for example, our one-switch control might waste fuel. Questions involving existence and synthesis of optimal controls can be quite difficult;

at this stage we will only intuitively discuss two special cases of the rocket car problem. We return to the complete problem in Chapters III and V.

The Time Optimal Problem ($\lambda_1 = 1, \lambda_2 = \lambda_3 = 0$)

Intuitively, it seems clear that we should use a bang–bang control in order to effect changes of state in minimum time, thus the synthesis developed above should be optimal. We will verify this in Chapter III.

The Minimum Fuel Problem ($\lambda_1 = \lambda_2 = 0, \lambda_3 = 1$)

Intuitively, it is easy to see that there should not be an optimal solution when $q(0) = 0$. This stems from the fact that if the rocket starts at rest, then we can fire an arbitrarily short burst and impart to the car an arbitrarily low velocity toward the origin (requiring a correspondingly small braking burst as we approach). Such a regime requires a very long time as the duration of the burst gets small, and there is no minimizing $u(\cdot)$ – we can consume as little fuel as we please by taking a long time to reach the target.

The reader might wish to verify the above analysis for initial state $(p_0, 0)$ with $p_0 < 0$ by setting $u_\alpha(t) = +1$ for $0 \le t \le \alpha$ and $u_\alpha(t) = -1$ for $|(p_0/\alpha)| \le t \le \alpha + |p_0/\alpha|$, with $u(t) \equiv 0$ otherwise. The associated cost is $C[u_\alpha(\cdot)] = 2\alpha$, so $\inf_{\mathscr{U}_1} C[u(\cdot)] = 0$. If $u_0(\cdot)$ were an optimal control, then $C[u_0(\cdot)] = 0$, which would imply $u(t) = 0$ a.e. But then the equation of motion of the rocket car, $\ddot{p}(t) = u(t)$, would imply that $p(t) \equiv p_0$, so such a control would not be successful.

The above discussion shows that we should keep a time-optimal part in the cost function $C[u(\cdot)]$ (i.e., take $\lambda_1 > 0$) to reject the long times required at low velocity. It is interesting to note that if the rocket is already moving *toward* the position $p = 0$, then there is an optimal control which minimizes fuel consumption (see Exercise 7). Finally, we should note that from a practical point of view, the above analysis does show how to keep fuel consumption low, depending on the time available.

Exercises

For Exercises 1 through 4, describe a possible interpretation of the words *system, state* and *control.* Describe a reasonable constraint set and target states.

1. A home heating system (thermostat plus furnace).

2. A boat with an outboard motor. The motor may be set with its screw propeller at a range of depths, it has a throttle, and can be turned so as to turn the boat. (Hint: the *state* might be thought of as position and (vector) velocity on the two-dimensional surface of a lake.)

3. An aircraft with one engine, ailerons on the back of the wings, horizontal elevators and a vertical rudder at the rear. The pilot has a throttle for the engine and controls for the deflections of the ailerons, elevators, and rudder. (Hint: the *state* of the aircraft might be a nine-dimensional vector: position (a three-vector), speed (a three-vector), rates of roll (r), yaw (y), and pitch (p).)

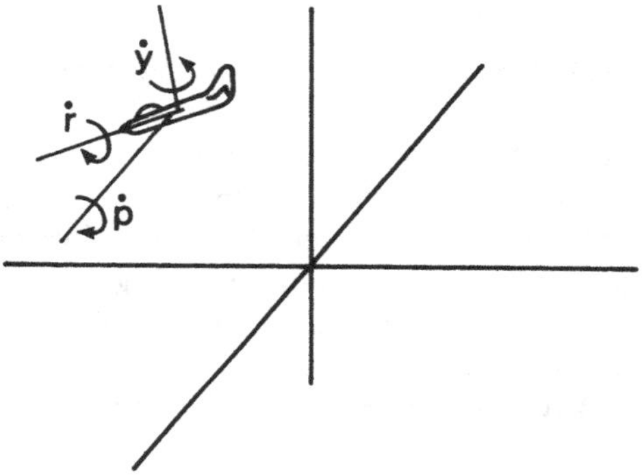

4. A tumor inside a human body. The tumor gets nourishment through the blood supply. It is attacked (controlled) by radiation and by chemicals in the blood supply. (Hint: the *state* might be described by: mass, density or volume, rate of growth or decrease.)

5. Sketch an example of the reachable set $K(t; \mathbf{x}_0)$ for the rocket car when the initial state is $\mathbf{x}_0 = (p_0, v_0)$ with $v_0 \neq 0$.

6. Describe $K(t; \mathbf{x}_0)$ and RC $(\mathbf{x}_0)$ for the one-dimensional problem

$$\dot{x}(t) = bx(t) + u(t), \qquad u(\cdot) \in \mathscr{U}_1, \qquad \mathscr{T}(t) \equiv 0,$$

for the two cases $b > 0$ and $b < 0$. Can you solve the least-time problem for $u(\cdot) \in \mathscr{U}_{PC}$?

7. Show that if the rocket car has initial state (p_0, v_0) with $p_0 > 0$, $v_0 < 0$, then there is an optimal control which minimizes fuel consumption, if $v_0^2 \leq 2p_0$. What happens if $v_0^2 > 2p_0$?

The following projects represent interesting areas of application of control theory. They can be treated as examples for perusal or exercises involving some outside reading.

8. *A Model for the Optimal Harvesting of Fish.* (*Reference*: *Clark* [1976].) The population $x(t)$ (measured as a mass in tons, for example) of a given type of fish is postulated to grow continuously in the absence of harvesting, according to the logistic equation.

$$\frac{dx}{dt} = rx\left(1 - \frac{x}{K}\right) \equiv F(x)$$

where the constant $r > 0$ is the net proportional growth rate (excess of birth rate over death rate), and the constant $K > 0$ is the carrying capacity of the environment (if $x(t) > K$, $dx/dt < 0$). With a *rate of harvesting* $h(t)$ (in tons per unit time, for example) the dynamics is described by

$$\frac{dx}{dt} = F(x) - h(t), \qquad x(0) = x_0.$$

Here $h(t)$ is the control. One simple assumed form for $h(t)$ might be $h(t) = E(t)x(t)$, where $E(t)$ is a combined "effort (by the fishermen) and catchability coefficient."

Show that if

(*) $\quad h(t) = Ex(t)$, where E is constant and $0 < E < r$,

then the system has an equilibrium at $x_1 = (K/r)(r - E)$. Thus we can maintain the population at x_1 under the harvest rate $Y = Ex_1 = KE(1 - E/r)$. Show also that each response $x(t; x_0, h(\cdot)) \to x_1$ as $t \to \infty$ (under (*) above). Make a graph of Y versus E for $0 < E < r$ (a yield-effort graph). Determine $x(t)$ explicitly when (*) holds.

A *cost* functional for this model (cf. p. 38 of Clark) is

$$C[h(\cdot)] = \int_0^\infty e^{-\delta t}\{p - c(x[t])x[t]\}h(t)\,dt,$$

where p is the (assumed constant) selling price of the fish per unit mass and $c(x[t])$ is the cost to catch a unit mass of the fish when the population is $x(t)$. The constant $\delta > 0$ is a fixed *discount rate* which can be roughly thought of as representing the interest lost (when the integration is carried out) if the fishermen do not harvest all of the fish immediately and put the profits in a savings account.

9. *A Model for the Control of Epidemics*. (*Reference*: Waltman [1974].) For models of certain diseases, one can divide the population into three disjoint classes of individuals;

$I(t)$: the infective class; those who can transmit the disease to others.
$S(t)$: the susceptible class; those who are not infective, but who are capable of contracting the disease and becoming infective.
$R(t)$: the removed class; constituting (a) those who have had the disease and are dead (for any reason), (b) those who have had the disease and are now immune, and (c) those who are completely isolated from the rest of the population until recovery with permanent immunity (or death).

Each of these three classes is expressed as a fraction, e.g., $S(t) = 0.5$ means that half the population is susceptible. The population is treated as a continuum for convenience, and the "progress" of an individual can be thought of as

$S \to I \to R$. The dynamics are assumed given by

$$\dot{S}(t) = -rS(t)I(t), \qquad S(0) = S_0 > 0,$$
$$\dot{I}(t) = rS(t)I(t) - \gamma I(t), \qquad I(0) = I_0 > 0,$$
$$\dot{R}(t) = \gamma I(t), \qquad R(0) = 0,$$

with $S_0 + I_0 = 1$, $S(t) + I(t) + R(t) \equiv 1$.

The constant $r > 0$ is called the infection rate; $\gamma > 0$ the removal rate, and $\rho = \gamma / r$ the relative removal rate (note that by the last equation, all removals from the population are assumed infected). One can use $R(t)$ as a control by introducing a quarantine, but a more realistic approach is to introduce a class $V(t)$ of vaccinated individuals and define $\alpha(t)$ to be the vaccination rate. The dynamics are then assumed to be of the form:

$$\dot{S}(t) = -rIS - \alpha, \qquad \dot{I}(t) = rIS - \gamma I$$
$$\dot{R}(t) = \gamma I, \qquad \dot{V}(t) = \alpha, \qquad V(0) = 0.$$

What are some practical restrictions on $\alpha(\cdot)$? What kind of cost function should be introduced? The following graph is a numerically obtained sketch of the progress of an epidemic, with and without vaccination. (From Figure 8.2 of Waltman.)

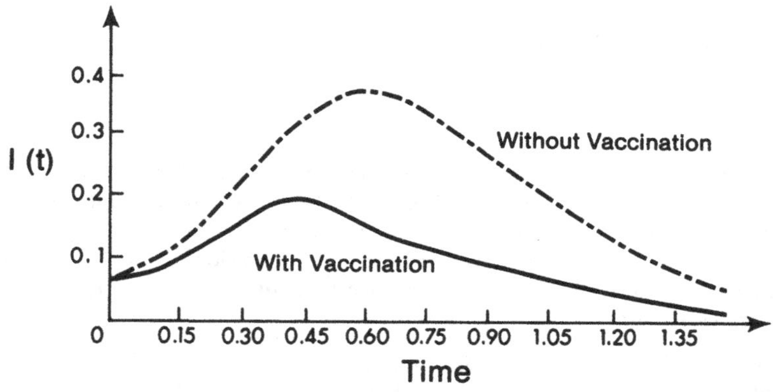

Among the many factors we have ignored in the above model are age dependence (e.g., elderly people are often more susceptible) and time delays (e.g., there is usually a time delay between exposure and infection). More complex models are presented in Frank Hoppensteadt's monograph, *Mathematical Theories of Populations: Demographics, Genetics and Epidemics*, Society for Industrial and Applied Math., Philadelphia, 1975.

10. *The Moon Landing Problem.* (*Reference*: *Fleming and Rishel* [1975].) We consider a simplified model for the problem of making a soft landing on the moon, using a minimum amount of fuel. Let m denote the mass of the spacecraft, and $h(t)$, $q(t)$ the height and vertical velocity of the craft relative to the surface of the moon. The control $u(t)$ will be the thrust of the spacecraft

engine. Taking the gravitational acceleration near the moon as a constant, γ, we can describe the dynamics:

$$\dot{h}(t)=q(t), \qquad h(0)=h_0,$$

$$\dot{q}(t)=-\gamma+\frac{1}{m}u(t), \qquad q(0)=q_0,$$

$$\dot{m}(t)=-ku(t), \qquad m(0)=m_0+F_0, \qquad 0\le u(t)\le 1,$$

where m_0 is the weight of the craft without fuel, and F_0 is the initial weight of the fuel on board. The target in (h, q, m)-space is

$$\mathcal{T}(t)=\{(0,0,z)|m_0\le z\le m_0+F_0\}.$$

Study the responses and reachable set for $u(\cdot)\in\mathcal{U}_{\text{BBPC}}$, with, for example, a single switch. If you have access to a large computer, check to see if it has a built in "Moon Landing Game."

11. *Neoclassical Economic Growth Model.* (*Reference*: *Intriligator* [1971].) The tools of the calculus of variations and optimal control theory have been used to analyze many dynamic questions in economics. An early application of these tools to the topic of optimal economic growth was by Ramsey [1928]. Hadley and Kemp [1971] have summarized more recent work using modern control theory and integrating "Ramsey's Problem" into neoclassical analysis of economic growth. Intriligator is a good source for a general introduction to the use of control theory and related topics in economic analysis and for the problem developed below. (Intriligator [1971], pp. 398–435.)

Let capital per unit of labor $k(t)$ be determined by the differential equation $\dot{k}(t)=sf(k(t))-nk(t)$ where $\dot{k}(t)$ represents the rate of change in capital per unit of labor. The marginal propensity to save s satisfies $0<s<1$. Labor grows at the constant proportional rate n. The aggregate production function determines output per unit of labor. It is assumed to be twice differentiable and to be a strictly concave, monotonic-increasing function satisfying the following conditions:

$$f'(k(t))>0, \qquad f''(k(t))<0,$$

$$f(0)=0, \qquad f(\infty)=\infty,$$

$$\lim_{k\to 0} f'(k(t))=\infty, \qquad \lim_{k\to\infty} f'(k(t))=0.$$

Utility per unit of labor is given by a function $U(c(t))$ where $c(t)$ is consumption per unit of labor obtained from $0<\{c(t)=(1-s)f(k(t))-\dot{k}(t)-nk(t)\}<\infty$. The utility function is twice differentiable with the assumptions $U'(c(t))>0$, $U''(c(t))<0$ representing positive but diminishing marginal utility. At different points in time utilities are assumed to be independent and additive. Present and future values of utility are compared by means of a discount factor $e^{-\delta t}$ where the discount rate is $\delta>0$.

The general optimal control problem is given by:

Maximize the welfare function

$$W=\int_0^T e^{-\delta t}U(c(t))\,dt$$

subject to

$$\dot{k}(t) = sf(k(t)) - nk(t),$$
$$c(t) = (1-s)f(k(t)) - \dot{k}(t) - k(t),$$
$$0 \leq c(t) \leq f(k(t)),$$
$$k(0) = k_0.$$

Solve this problem under each of the following conditions:

(a) $T \to \infty$,
(b) $T = t_1$ and $k(t_1) \geq k_1$,
(c) Re-do (a) and (b) but let $U(c(t)) = c(t)$.

Notes for Chapter I

1. We have described only a few situations which can be modelled as control problems. For additional material see Clark [1976] for management of renewable resources, Milhorn [1966] for applications to physiology, and Auslander, Takahashi, and Rabins [1974] for a wide range of models. The last reference has (on pp. 203–211) an analysis of the water-level regulator described as Example 2. For a historical discussion of feedback control, see Mayr [1970], and Bennett [1979].

2. *Differential Equations.* A summary of basic results can be found in the Mathematical Appendix.

Existence and uniqueness results for the solution(s) of the initial value problem

$$\dot{\mathbf{x}} = \mathbf{f}(t, \mathbf{x}), \qquad \mathbf{x}(t_0) = \mathbf{x}_0, \qquad t \in I, \qquad \mathbf{x}(t) \in R^n,$$

with I an interval containing t_0, can be found in any advanced textbook on ordinary differential equations, for example, Chapters I and III of Coddington and Levinson [1955] or Chapters II and III of Hartman [1964]. Extendability of solutions is still a subject of research – the techniques used and results obtained depend very much on the particular form of the equation. Surveys of some techniques and results are given in Bellman [1953], Cesari [1963], Coppel [1965], Sansone and Conti [1964].

3. *Game Theory.* In certain areas of game theory one can find problems similar to our general control problem, but with the important exception that the behavior of the target $\mathscr{T}(t)$ as a function of time is not known to us in advance – there is an opponent capable of influencing $\mathscr{T}(t)$, who is trying to keep us from hitting the target. In other game theory problems, the opponent might be able to directly affect our state $\mathbf{x}(t; \mathbf{x}_0, \mathbf{u}(\cdot))$ and/or the control $\mathbf{u}(\cdot)$ which we have chosen. We will not treat those types of problems here. For a complete discussion of the basic theory of differential

games, see Isaacs [1975]. On pp. 365–368, Isaacs shows how differential game theory can be applied to control theory.

4. *The Calculus of Variations.* Optimal control theory is intimately connected with the calculus of variations. We will give one simple example to illustrate this connection; for a thorough treatment see Hestenes [1966], and the survey by McShane [1978].

We use a simple form of the problem of Bolza from the calculus of variations. Let $t_0 < t_1$ be given real numbers, and let $\mathcal{S}$ be the class of functions $x(\cdot): [t_0, t_1] \to R$, absolutely continuous on $[t_0, t_1]$, that satisfy $x(t_0) = 0$, $x(t_1) = 1$ and are such that

$$J(x) = \int_{t_0}^{t_1} f^0(t, x(t), \dot{x}(t))\, dt \quad \text{exists,}$$

where $f^0: [t_0, t_1] \times R \times R \to R$ is a given continuous function. The problem of Bolza is to determine $\lambda = \inf_{\mathcal{S}} J(x)$ and to find $x_*(\cdot) \in \mathcal{S}$ that yields this minimum value (when λ and $x_*(\cdot)$ exist). To restate this problem as a control problem, we replace $\dot{x}(t)$ by a control $u(t)$, and specify $\dot{x}(t) = u(t)$ as the dynamics of our system. Therefore, our optimal control problem is to determine $u(\cdot)$ so that the *cost* function

$$C[u(\cdot)] = \int_{t_0}^{t_1} f^0(t, x(t), u(t))\, dt$$

is minimized for the control problem:

$$\dot{x}(t) = u(t), \qquad x(t_0) = 0.$$

The constraint on $u(\cdot)$ is:

$$\int_{t_0}^{t_1} u(t)\, dt = 1,$$

which forces $x(t_1) = 1$. Thus the target state $\mathcal{T}(t) \equiv 1$ is automatically hit because of the constraint on $u(\cdot)$. Alternatively, we could remove the constraint on $u(\cdot)$ and specify $x(t_1) = 1$ (note that the time of hitting the target is also specified).

Conversely, one can under certain conditions interpret a control problem as a Bolza problem. For a description of this interpretation, see the book of Hestenes referred to above, or pp. 34–38 of Berkovitz [1974].

5. *Dynamic Programming.* Dynamic programming was originally conceived by R. Bellman as an effective computational method for dealing with optimal decision making in discrete-time processes. Later it was recognized that the same principles can be used in the treatment of continuous-time processes, including problems in the calculus of variations

and optimal control theory. A full description of the basic ideas can be found in Chapter 5 of Bellman [1967]. A thorough, rigorous treatment of the use of dynamic programming in optimal control theory can be found in Chapter IV of Fleming and Rishel [1975]. In Chapter V we give a simple dynamic programming proof of a restricted form of the Pontryagin Maximum Principle, and the relevant theory is developed there.

6. *Stochastic Effects.* The dynamics of a system might be affected by some random phenomena, for example "white noise" in communications theory or "random genetic drift" in genetics. The dynamics might then be described by a vector *stochastic* ordinary differential equation

$$\dot{\mathbf{x}}(t) = \mathbf{f}(t, \mathbf{x}, \mathbf{u}, \boldsymbol{\xi}(\cdot))$$

where $\boldsymbol{\xi}(\cdot)$ is a probability distribution. The response $\mathbf{x}(\cdot)$ must then be interpreted as an evolving probability distribution. For a thorough treatment of this area, see Fleming and Rishel [1975], and Balakrishnan [1973].

7. *Observables, Time Lags, Functional Equations.* Sometimes we pick a state variable $\mathbf{x}(t)$ that cannot be observed directly – perhaps only certain components of $\mathbf{x}(t)$, or some function of $\mathbf{x}(t)$, is physically measurable. A simple example would be an electronic circuit containing micro-electronic devices, for which one of the state variable components, $x^i(t)$, is the current in some part of a microscopic chip. These chips are so small that one cannot usually make current measurements in their constituent parts. Engineers have a simple way of distinguishing the *state* $\mathbf{x}(t)$ of a system (which may not be observable) from its observable *output* $\mathbf{c}(t)$ (something we can measure) by using a picture:

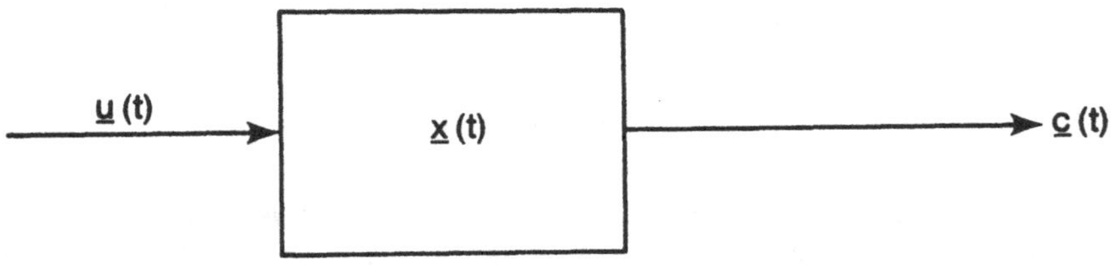

The state is inside the box, "out of sight." We can only "see" $\mathbf{c}(t)$. Mathematically, we then must deal with the behavior of $\mathbf{c}(t) = \mathbf{F}(\mathbf{x}(t), t)$, where $\mathbf{F}$ is assumed known. In many cases, there is a matrix H such that $\mathbf{c}(t) = H\mathbf{x}(t)$. For a discussion of observability results, see Lee and Markus [1967].

There is also the question of time lags. In most real-world situations, reactions are not instantaneous. This means that the dynamics of a system may involve one or more time lags, for example,

$$\dot{\mathbf{x}}(t) = \mathbf{f}(t, \mathbf{x}(t), \mathbf{x}(t-\tau), \mathbf{u}(t-\tau)),$$

or the system might in fact have a memory:

$$\mathbf{x}(t) = \int_{t_0}^{t} M(t, s)\mathbf{f}(s, \mathbf{x}(s), \mathbf{u}(s))\, ds.$$

This is presently an active area of research. The book of M. N. Oğuztöreli [1966] treats this type of problem.

Chapter II

Controllability

1. Introduction: Some Simple General Results

The Mathematical Appendix at the end of the book contains most of the basic results from the areas of linear algebra, functional analysis, convexity, and differential equations which we will need in this chapter.

Recall that $\mathscr{C}$ is the controllable set, i.e., the set of all initial points $\mathbf{x}_0$ which can be steered to the target. We will prove some general results on the nature of the controllable set $\mathscr{C}$ for both the general autonomous problem

(NLA) $$\dot{\mathbf{x}} = \mathbf{f}(\mathbf{x}, \mathbf{u}), \qquad \mathbf{x}(t) \in R^n, \qquad \mathbf{u}(\cdot) \in \mathscr{U}_m,$$

and the linear autonomous problem

(LA) $$\dot{\mathbf{x}} = A\mathbf{x} + B\mathbf{u}, \quad A, B \text{ constant matrices,}$$

with target $\mathscr{T}(t) \equiv \mathbf{0}$. For simplicity we assume that $\mathbf{f}(\mathbf{0}, \mathbf{0}) = \mathbf{0}$ and that $\mathbf{f}(\mathbf{x}, \mathbf{u})$ is continuously differentiable on $R^n \times R^m$. Then for a given initial state $\mathbf{x}_0$, the response to a given control is unique. Because both (NLA) and (LA) are autonomous, if $\mathbf{u}(t)$ steers $\mathbf{x}_0$ to $\mathbf{0}$ on $[0, t_1]$ with response $\mathbf{x}[t]$, then $\mathbf{u}(t - t_0)$ steers $\mathbf{x}_0$ to $\mathbf{0}$ on $[t_0, t_1 + t_0]$ with response $\mathbf{x}[t - t_0]$. Therefore our standard assumption that $t_0 = 0$ involves no loss of generality. Also, our assumption $\mathscr{T}(t) \equiv \mathbf{0}$ is for simplicity; any fixed target state $\mathbf{x}_1$ would do.

Our main results for (LA) are as follows:

In R^n, the set $\mathscr{C}$ is arcwise connected, symmetric, and convex. Also,

$$\mathscr{C} \textit{ is open} \Leftrightarrow \textit{the target } \mathbf{0} \in \operatorname{Int} \mathscr{C} \Leftrightarrow \operatorname{rank} M = n,$$

where the controllability matrix M is defined as

$$M = \{B, AB, \ldots, A^{n-1}B\},$$

and "Int" *means interior. Finally,* $\mathscr{C} = R^n$ *if and only if* $\operatorname{rank} M = n$ *and no eigenvalue of A has positive real part.*

For (NLA), we form the Taylor expansion $\mathbf{f}(\mathbf{x}, \mathbf{u}) = A\mathbf{x} + B\mathbf{u} + o(|\mathbf{x}| + |\mathbf{u}|)$, where $a_{ij} = \partial f^i/\partial x^j$, $b_{ik} = \partial f^i/\partial u^k$, both evaluated at $\mathbf{x} = \mathbf{0}$, $\mathbf{u} = \mathbf{0}$, $i, j = 1, \ldots, n$; $k = 1, \ldots, m$. We define $M_f = \{B, AB, \ldots, A^{n-1}B\}$.

Then for (*NLA*)

$$\operatorname{rank} M_f = n \Rightarrow \mathscr{C} \text{ is open,}$$

but the converse is false. If $\operatorname{rank} M_f = n$ *and the trivial solution* $\mathbf{x}[t] \equiv 0$ *of* $\dot{\mathbf{x}} \equiv \mathbf{f}(\mathbf{x}, \mathbf{0})$ *is globally asymptotically stable, then* $\mathscr{C} = R^n$.

Finally, we will show that some (but not all) of these results hold if, instead of being able to use any measurable control from $\mathscr{U}_m$, we are restricted to special control classes: piecewise constant, or smooth, or bang–bang controls.

For $\mathbf{x}_0 \in \mathscr{C}$, Figure 1 illustrates a typical response path which starts at $\mathbf{x}_0 \equiv \mathbf{x}[0]$ and ends at the target $\mathbf{0} = \mathbf{x}[t_1]$. We will now show that for (NLA),

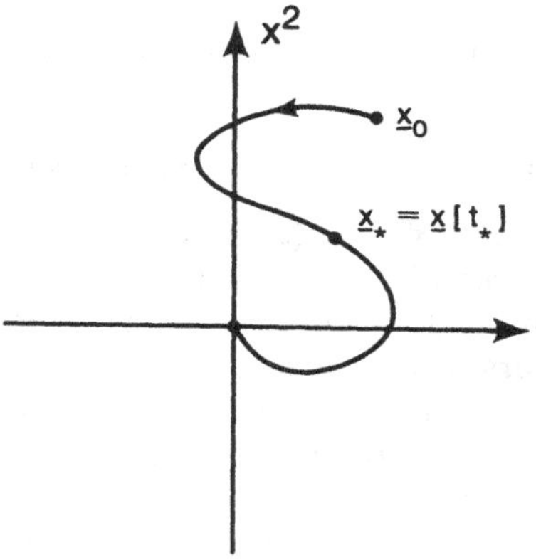

Figure 1

any point $\mathbf{x}_* = \mathbf{x}[t_*]$ on this path is also in $\mathscr{C}$. That is, there is a control $\tilde{\mathbf{u}}(\cdot)$ such that the solution of

$$\dot{\mathbf{x}} = \mathbf{f}(\mathbf{x}, \tilde{\mathbf{u}}), \qquad \mathbf{x}[0] = \mathbf{x}_*$$

arrives at the target $\mathbf{0}$ at some time t_2. This result does *not* hold in general for systems in which t appears explicitly (non-autonomous systems). The reader is asked to work out an example in Exercise 3 at the end of this chapter.

Suppose $\mathbf{x}_0 \in \mathscr{C}$ and let $\mathbf{x}[t] = \mathbf{x}(t; \mathbf{x}_0, \mathbf{u}(\cdot))$ be the response to the successful control $\mathbf{u}(\cdot)$, with $\mathbf{x}[t_1] = \mathbf{0}$. Let $t_* \in [0, t_1]$: we must show that the point $\mathbf{x}_* = \mathbf{x}[t_*]$ also belongs to $\mathscr{C}$. Because the system is autonomous, the following control is successful:

$$\tilde{\mathbf{u}}(t) = \mathbf{u}(t + t_*), \qquad 0 \le t \le t_1 - t_*,$$

with response $\tilde{\mathbf{x}}[t] \equiv \mathbf{x}[t + t_*]$. To see this, we write

$$\frac{d}{dt}\tilde{\mathbf{x}}[t] = \frac{d}{dt}\mathbf{x}[t + t_*] = \mathbf{f}(\mathbf{x}(t + t_*), \mathbf{u}(t + t_*)) \equiv \mathbf{f}(\tilde{\mathbf{x}}[t], \tilde{\mathbf{u}}(t)),$$

$$\tilde{x}[0] = x[t_*] = \mathbf{x}_*, \qquad \tilde{\mathbf{x}}[t_1 - t_*] = \mathbf{x}[t_1 - t_* + t_*] = \mathbf{x}[t_1] = \mathbf{0}.$$

Therefore, $\mathbf{x}_* \in \mathscr{C}$.

Before we discuss general controllability problems, we remind the reader that $\mathscr{U}_m$ is the class of measurable functions, $\mathbf{u}(\cdot)$, each defined on some interval $[0, t_1]$ (t_1 may depend on $\mathbf{u}(\cdot)$), with values $\mathbf{u}(t) \in \Omega$, where Ω is the unit cube in R^m. The controllable set is defined by $\mathscr{C} \equiv \bigcup_{t_1 > 0} \mathscr{C}(t_1)$, where

$$\mathscr{C}(t_1) = \{\mathbf{x}_0 \in R^n \mid \exists \mathbf{u}(\cdot) \in \mathscr{U}_m \text{ such that } \mathbf{x}(t_1; \mathbf{x}_0, \mathbf{u}(\cdot)) = \mathbf{0}\}.$$

Because $\mathbf{f}(\mathbf{0}, \mathbf{0}) = \mathbf{0}$, the target $\mathbf{0}$ is in $\mathscr{C}$. (If $\mathbf{x}_0 = \mathbf{0}$, then $\mathbf{u}(t) \equiv \mathbf{0}$ is successful, with response $\mathbf{x}[t] \equiv 0$.)

The basic questions in the area of controllability are

(1) to describe $\mathscr{C}$,
(2) to show how $\mathscr{C}$ is changed if we use special classes of controls.

Two desirable properties of $\mathscr{C}$ are: (a) $\mathbf{0} \in \operatorname{Int} \mathscr{C}$ and (b) $\mathscr{C} = R^n$. In case (b) the system is called *completely controllable.*

The reason for wanting to know conditions under which $\mathscr{C} = R^n$ is obvious. There are several good reasons for wanting $\mathbf{0} \in \operatorname{Int} \mathscr{C}$, that is, for wanting some open ball about $\mathbf{0}$, of radius δ, inside $\mathscr{C}$: $\mathscr{B}(\mathbf{0}; \delta) \subset \mathscr{C}$. First, it is nice to know that we can steer to the target if we are close enough to begin with (regardless of direction). Second, a mathematical model is of course an idealization of some real-world system, and several (hopefully) small effects may have been neglected. These facts, plus our inability to carry out measurements with arbitrarily fine precision, force us to realize that even if we reach the target, there may be all kinds of effects tending to move us slightly off. It is reassuring to know that if these perturbations are small enough, we can always steer back to the target. It is also desirable that $\mathscr{C}$ be open since real-world data concerning the initial state cannot be exact.

Any results about $\mathscr{C}$ or $\mathscr{C}(t_1)$ which apply to *all* equations of the form (NLA) can be restated in terms of reachable sets. Notice that $\mathbf{x}(t)$ solves (NLA) with $\mathbf{x}(0) = \mathbf{x}_0$, and $\mathbf{x}(t_1) = \mathbf{x}_1$, if and only if $\mathbf{z}(t) = \mathbf{x}(t_1 - t)$ solves the *time-reversed* system

$$(*) \qquad \dot{\mathbf{z}} = -f(\mathbf{z}, \tilde{\mathbf{u}}), \qquad \mathbf{z}(0) = \mathbf{x}_1, \qquad \mathbf{z}(t_1) = \mathbf{x}_0, \qquad \tilde{\mathbf{u}}(t) = \mathbf{u}(t_1 - t).$$

Notice that $(*)$ is again a non linear autonomous equation. The two systems have the same curves as trajectories, traversed in opposite directions (Figure 2).

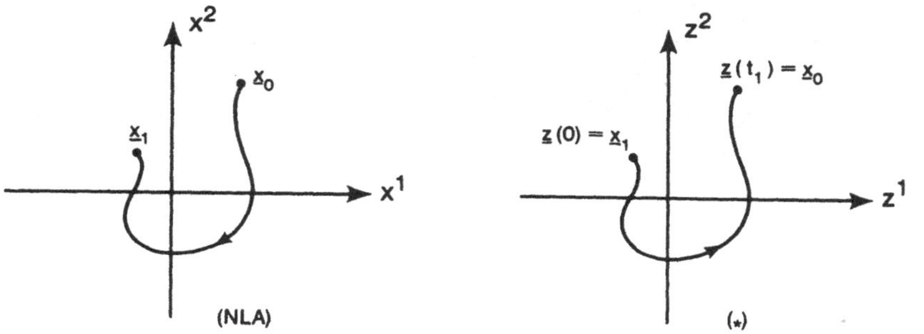

Figure 2

This means that the set $\mathscr{C}(t_1)$ of those states which can be steered to $\mathbf{x}_1$ at time t_1 under (NLA) is the same as the reachable set $K^{-}(t_1; \mathbf{x}_1)$ (those states which can be *reached* at time t_1 from $\mathbf{x}_1$) *for the time-reversed system.* Thus when we assert "$\mathscr{C}(t_1)$ is symmetric for *every* equation (NLA)" we are also asserting "$K(t_1, \mathbf{x}_1)$ is always symmetric."

Recall that a set in R^n is *arcwise connected* if any two points in the set can be joined by a path $\mathbf{p}(t)$, which never leaves the set.

Theorem 1. *For the system (NLA) described above, $\mathscr{C}$ is arcwise connected. $\mathscr{C}$ is open if and only if $\mathbf{0} \in \operatorname{Int} \mathscr{C}$.*

Proof. First notice that if $\mathbf{x}_0 \in \mathscr{C}$, then there is a path joining $\mathbf{x}_0$ to $\mathbf{0}$, namely the successful response $\mathbf{x}(t; \mathbf{x}_0, \mathbf{u}(\cdot))$. Furthermore, all points on this path belong to $\mathscr{C}$, as remarked earlier. If $\mathbf{x}_0$, $\tilde{\mathbf{x}}_0$ are both in $\mathscr{C}$, then they are joined to $\mathbf{0}$ by response paths, thus they are joined to each other by a path consisting of points from $\mathscr{C}$.

If $\mathscr{C}$ is open, then since $\mathbf{0} \in \mathscr{C}$ we conclude $\mathbf{0} \in \operatorname{Int} \mathscr{C}$.

Now suppose that $\mathbf{0} \in \operatorname{Int} \mathscr{C}$ so there exists an open ball $\mathscr{B}_0 \equiv \mathscr{B}(\mathbf{0}, \delta_0) \subset \mathscr{C}$. Let $\mathbf{x}_* \in \mathscr{C}$. We want to show that there is an open ball $\mathscr{B}(\mathbf{x}_*; \delta) \subset \mathscr{C}$. Now there exists a successful control $\mathbf{u}_*(t)$ that steers $\mathbf{x}_*$ to $\mathbf{0}$, that is, the solution of $\dot{\mathbf{x}} = \mathbf{f}(\mathbf{x}, \mathbf{u}_*)$, $\mathbf{x}(0) = \mathbf{x}_*$ satisfies $\mathbf{x}(t_1; \mathbf{x}_*, \mathbf{u}_*(\cdot)) = \mathbf{0}$ for some $t_1 > 0$. Since $\mathbf{f}(\mathbf{x}, \mathbf{u})$ is differentiable, the solutions of (NLA) depend continuously on the initial value of $\mathbf{x}(0)$. Thus, for the fixed control $\mathbf{u}_*(\cdot)$, and the ball $\mathscr{B}_0$, there is a ball $\mathscr{B}_* = \mathscr{B}(\mathbf{x}_*; \delta_*)$ such that (Figure 3)

$$\mathbf{x}[0] = \mathbf{x}_0 \in \mathscr{B}_* \Rightarrow \mathbf{x}[t_1] \in \mathscr{B}_0 \subset \mathscr{C}.$$

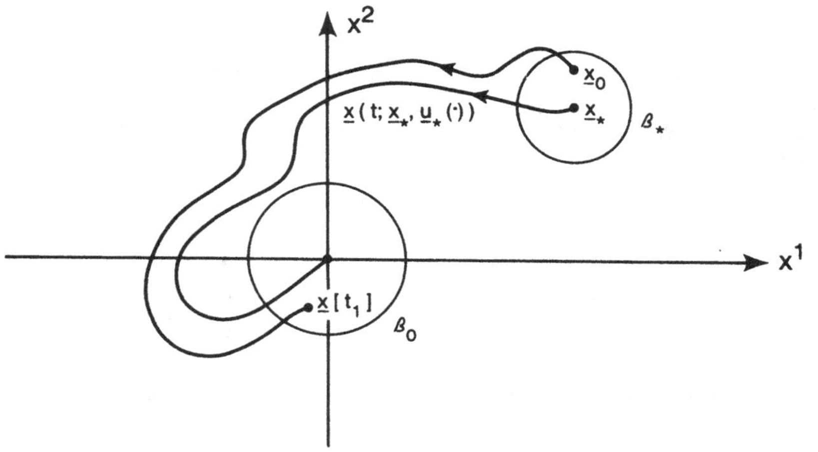

Figure 3

Since $\mathbf{x}[t_1]$ is in $\mathscr{C}$, there is a control $\tilde{\mathbf{u}}(t)$ that steers the point $\mathbf{x}[t_1]$ to $\mathbf{0}$ over the time interval $[0, t_2]$. Then $\mathbf{u}(\cdot)$ defined by $\mathbf{u}(t) = \mathbf{u}_*(t)$ for $0 \leq t \leq t_1$, $\mathbf{u}(t) = \tilde{\mathbf{u}}(t - t_1)$ for $t_1 < t \leq t_1 + t_2$, steers $\mathbf{x}_0$ to $\mathbf{0}$. Thus $\mathscr{B}_* \subset \mathscr{C}$. □

2. The Linear Case

To investigate $\mathscr{C}$ more deeply, it is worthwhile to investigate a system whose dynamics is described by a *linear* autonomous vector differential equation. We can then use the technique of linearization to study certain nonlinear cases.

Therefore, we will study controllability problems for the linear autonomous system:

(LA) $$\dot{\mathbf{x}} = A\mathbf{x} + B\mathbf{u}, \quad A, B \text{ constant matrices.}$$

By the variation of parameters formula (cf. The Mathematical Appendix), for a given $\mathbf{u}(\cdot) \in \mathscr{U}_m$ the solution of (LA) with initial state $\mathbf{x}_0$ at time $t = 0$ is given by the *response formula*

(1) $$\mathbf{x}[t] \equiv \mathbf{x}(t; \mathbf{x}_0, \mathbf{u}(\cdot)) = X(t)X^{-1}(0)\mathbf{x}_0 + \int_0^t X(t)X^{-1}(s)B(s)\mathbf{u}(s)\,ds$$

where $X(t)$ is any fundamental matrix for the homogeneous equation $\dot{x}(t) = A\mathbf{x}(t)$. In particular, $\mathbf{x}_0 \in \mathscr{C}(t_1)$ (we hit the target $\mathscr{T}(t) \equiv \mathbf{0}$ at time t_1) if and only if there is a $\mathbf{u}(\cdot) \in \mathscr{U}_m$ such that

(*) $$X(t_1)X^{-1}(0)\mathbf{x}_0 + \int_0^{t_1} X(t_1)X^{-1}(s)B(s)\mathbf{u}(s)\,ds = \mathbf{0}.$$

Theorem 2. *For the system (LA), the controllable set $\mathscr{C} \subset R^n$ is symmetric and convex.*

Proof. Manipulating (*) we see that $\mathbf{x}_0 \in \mathscr{C}(t_1)$ if and only if there is a $t_1 > 0$ and a $\mathbf{u}(\cdot) \in \mathscr{U}_m[0, t_1]$ such that

$$\mathbf{x}_0 = -X(0) \int_0^{t_1} X^{-1}(s)B(s)\mathbf{u}(s)\, ds. \tag{2}$$

By (2), if $\mathbf{x}_0 \in \mathscr{C}(t_1)$ (using $\mathbf{u}(\cdot)$), then $-\mathbf{x}_0 \in \mathscr{C}(t_1)$ (using $-\mathbf{u}(\cdot)$), so $\mathscr{C} = \bigcup_{t_1>0} \mathscr{C}(t_1)$ is symmetric.

Since integration is linear, and $\mathscr{U}_m[0, t_1]$ is convex, it easily follows from (2) that $\mathscr{C}(t_1)$ is convex: if $\mathbf{x}_0 \in \mathscr{C}(t_1)$ with control $\mathbf{u}_0(\cdot)$ and $\mathbf{x}_* \in \mathscr{C}(t_1)$ with control $\mathbf{u}_*(\cdot)$, then $[\alpha\mathbf{x}_0 + (1-\alpha)\mathbf{x}_*] \in \mathscr{C}(t_1)$ with control $[\alpha\mathbf{u}_0(\cdot) + (1-\alpha)\mathbf{u}_*(\cdot)]$. But the union $\mathscr{C} = \bigcup_{t_1>0} \mathscr{C}(t_1)$ of convex sets may not be convex. To show that $\mathscr{C}$ is in fact convex, let $\mathbf{x}_0 \in \mathscr{C}(t_1)$, $\mathbf{x}_* \in \mathscr{C}(t_*)$. Then (2) holds for $\mathbf{x}_0$, and $\mathbf{x}_*$ satisfies

$$\mathbf{x}_* = -X(0) \int_0^{t_*} X^{-1}(s)B(s)\mathbf{u}_*(s)\, ds.$$

Without loss of generality, assume $t_1 < t_*$.

If we define a new control $\mathbf{u}_0(t)$ to be equal to $\mathbf{u}(t)$ from (2) on $[0, t_1]$, and $\mathbf{0}$ on $(t_1, t_*]$, then (2) can be written with the upper limit t_*:

$$\mathbf{x}_0 = -X(0) \int_0^{t_*} X^{-1}(s)B(s)\mathbf{u}_0(s)\, ds,$$

which means $\mathbf{x}_0 \in \mathscr{C}(t_*)$. Since $\mathscr{C}(t_*)$ is convex, any convex combination of $\mathbf{x}_*$ and $\mathbf{x}_0$ belongs to $\mathscr{C}(t_*) \subset \mathscr{C}$. □

Remarks

1. The above theorem holds (with identical proof) for $A(t)$, $B(t)$ continuous, not necessarily constant, matrices.
2. The above theorem is valid for any control class $\mathscr{U}$ which is both symmetric and convex. Therefore, this theorem holds for the classes $\mathscr{U}_{\mathrm{PC}}$ (piecewise constant controls) and $\mathscr{U}_\varepsilon$ (smooth controls). We denote the corresponding controllable sets by $\mathscr{C}_{\mathrm{PC}}$ and $\mathscr{C}_\varepsilon$ respectively.
3. The above proof will not work for the classes of bang–bang controls $\mathscr{U}_{\mathrm{BB}}[0, t_1]$ and $\mathscr{U}_{\mathrm{BBPC}}[0, t_1]$, since they are symmetric but not convex; the proof also used a control which takes the value $\mathbf{0}$, which is thus not in $\mathscr{U}_{\mathrm{BB}}$. We denote the controllable sets for $\mathscr{U}_{\mathrm{BB}}[0, t_1]$, $\mathscr{U}_{\mathrm{BBPC}}[0, t_1]$ by $\mathscr{C}_{\mathrm{BB}}(t_1)$, $\mathscr{C}_{\mathrm{BBPC}}(t_1)$, respectively. In the Appendix to this chapter we will prove:

   For the linear autonomous system (LA), $\mathscr{C}_{\mathrm{BB}}(t_1)$ coincides with $\mathscr{C}(t_1)$; this set is compact, convex, and depends continuously (in the Hausdorff metric) on t_1.

This implies that for (LA), $\mathscr{C} \equiv \bigcup_{t_1>0} \mathscr{C}(t_1) = \bigcup_{t_1>0} \mathscr{C}_{BB}(t_1) \equiv \mathscr{C}_{BB}$. Since $\mathscr{C}$ is convex, it follows that $\mathscr{C}_{BB}$ is convex. In fact the following surprising result holds (see Lee and Markus [1967], pp. 155–168).

The Bang–Bang Principle for

$$\dot{\mathbf{x}}(t) = A(t)\mathbf{x} + B(t)\mathbf{u} + \mathbf{b}(t)$$

For the linear system above, suppose that our usual assumptions hold, except that the values of $\mathbf{u}(t)$ are only constrained to lie in some arbitrary fixed compact set $\Psi \subset R^m$. Then the corresponding controllable set $\mathscr{C}_\Psi(t_1)$ is compact, convex, and depends continuously on t_1. Moreover, if Ψ, Ψ_* are compact subsets of R^m with the same convex hull, then the corresponding controllable sets coincide.

Bang–bang controls take their values ($|u^i(t)| = 1$) on the vertices of the unit cube. Since the vertices of the unit cube $\Omega \subset R^m$ (or in general, the extreme points of any given convex set) have the same convex hull as Ω (namely Ω itself) it follows from the above theorem that $\mathscr{C}_{BB}(t_1) = \mathscr{C}(t_1)$ for all $t_1 > 0$; thus the theorem we prove in the Appendix is a special case of the above result.

Notice that the above proposition says nothing about the convexity of the controllable set $\mathscr{C}_\Psi \equiv \bigcup_{t_1>0} \mathscr{C}_\Psi(t_1)$. If we use Ψ_* to denote the convex hull of Ψ, and if $\mathbf{0}$ belongs to this convex hull, then in fact $\mathscr{C}_\Psi$ is convex. To see this, note that $\mathscr{C}_{\Psi_*}$ is convex by the same proof as that of Theorem 2, and also $\mathscr{C}_{\Psi_*}(t_1) = \mathscr{C}_\Psi(t_1)$ for all t_1 by the Bang–Bang Principle. Thus

$$\mathscr{C}_\Psi \equiv \bigcup_{t_1>0} \mathscr{C}_\Psi(t_1) = \bigcup_{t_1>0} \mathscr{C}_{\Psi_*}(t_1) \equiv \mathscr{C}_{\Psi_*},$$

so $\mathscr{C}_\Psi$ is in fact convex.

Example 1 ($\mathscr{C}$ May Not Contain a Neighborhood of the Target). Consider the two-dimensional system $\dot{p} = p + u$, $\dot{q} = q + u$, $-1 \le u(t) \le 1$, which is a linear system with

$$\mathbf{x}(t) = \begin{bmatrix} p(t) \\ q(t) \end{bmatrix}, \qquad A = \begin{bmatrix} 1 & 0 \\ 0 & 1 \end{bmatrix}, \qquad B = \begin{bmatrix} 1 \\ 1 \end{bmatrix},$$

$$n = 2, \qquad m = 1, \qquad \mathscr{T}(t) \equiv \mathbf{0}.$$

The solution that satisfies the initial condition $\mathbf{x}(0) = \mathbf{x}_0$ is

$$\mathbf{x}(t) = e^t \mathbf{x}_0 + \left\{ e^t \int_0^t e^{-s} u(s)\, ds \right\} \begin{bmatrix} 1 \\ 1 \end{bmatrix}.$$

If $p(0) > 1$ and $q(0) > 1$, then the differential equations imply $\dot{p}(t) > 0$, $\dot{q}(t) > 0$ for all t. Clearly $\mathscr{C} = \{(p, q) | p = q,\ |p| < 1,\ |q| < 1\}$, so $\mathscr{C}$ does not contain a neighborhood of the origin (Figure 4). Dropping all restrictions on the range of $u(\cdot)$ will not change this conclusion.

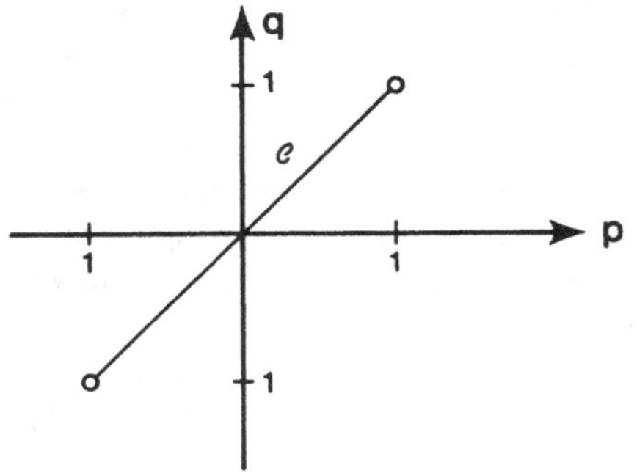

Figure 4

The above example shows that we need some restrictions on the matrices A and B in the system (LA) to ensure that $\mathscr{C}$ contains a neighborhood of our target, $\mathbf{0}$. The key to this problem is the $n \times (mn)$ *controllability matrix* $M(A, B)$:

$$M \equiv [B, AB, A^2B, \ldots, A^{n-1}B].$$

Theorem 3. *For the linear autonomous system* (*LA*),

$$\operatorname{rank} M = n \Leftrightarrow \mathbf{0} \in \operatorname{Int} \mathscr{C} (\Leftrightarrow \mathscr{C} \textit{ is open, by Theorem } 1).$$

Proof. We will actually prove the equivalent statement $\mathbf{0} \notin \operatorname{Int} \mathscr{C} \Leftrightarrow \operatorname{rank} M < n$.

For $t_1 > 0$, we recall from (2) that $\mathbf{x}_0 \in \mathscr{C}(t_1)$ if and only if there is $\mathbf{u}(\cdot) \in \mathscr{U}_m$, such that

$$\mathbf{x}_0 = -\int_0^{t_1} e^{-As} B\mathbf{u}(s)\, ds, \tag{3}$$

using the fundamental matrix $X(t) = e^{At}$ for (LA).

First, suppose $\operatorname{rank} M < n$. Then there is a unit vector $\mathbf{y} \in R^n$, $\|\mathbf{y}\| = 1$, perpendicular to every column of M, that is, the $1 \times m$ row vector $\mathbf{y}^T A^k B = \mathbf{0}$ for $k = 0, 1, \ldots, (n-1)$. If $\mathscr{P}(\lambda) \equiv \det(\lambda I - A)$ is the characteristic polynomial of A, then the Cayley–Hamilton Theorem implies that $\mathscr{P}(A) = 0$. This in turn implies that A^n can be written as a linear combination of the lower powers, $A^n = \beta_1 A^{n-1} + \cdots + \beta_n (*)$, and so

$$\mathbf{y}^T A^n B = \beta_1 \mathbf{y}^T A^{n-1} B + \beta_2 \mathbf{y}^T A^{n-2} B + \cdots + \beta_n \mathbf{y}^T B = \mathbf{0}.$$

Multiplying (*) by $\mathbf{y}^T A$ then gives $\mathbf{y}^T A^{n+1} B = \mathbf{0}$, and continuing in this way, we see that $\mathbf{y}^T A^k B = \mathbf{0}$ for $k = 0, 1, 2, \ldots$. But

$$e^{-As} = \sum_{k=0}^{\infty} \frac{(-1)^k A^k}{k!} s^k,$$

so $\mathbf{y}^T e^{-As}B = \mathbf{0}$. Thus for all $\mathbf{x}_0 \in \mathscr{C}(t_1)$, $\mathbf{y}^T\mathbf{x}_0 = 0$ by (3), and so $\mathscr{C}(t_1)$ lies in the hyperplane perpendicular to $\mathbf{y}$, *for all* $t_1 > 0$. Since $\mathscr{C}$ is the union of the sets $\mathscr{C}(t_1)$, it lies in the same hyperplane, and $\mathbf{0} \notin \text{Int } \mathscr{C}$.

Conversely, suppose $\mathbf{0} \notin \text{Int } \mathscr{C}$. Then for all $t_1 > 0$, $\mathbf{0} \notin \text{Int } \mathscr{C}(t_1)$, since $\mathscr{C}(t_1) \subset \mathscr{C}$. Now $\mathbf{0} \in \mathscr{C}(t_1)$ (set $\mathbf{u}(\cdot) \equiv \mathbf{0}$), for all t_1, and $\mathscr{C}(t_1)$ is convex, so for each t_1 there must exist a hyperplane through $\mathbf{0}$ such that $\mathscr{C}(t_1)$ lies on one side of this hyperplane, i.e., there is a non-trivial vector $\mathbf{b}(t_1)$ such that for all $\mathbf{x}_0 \in \mathscr{C}(t_1)$, $\mathbf{b}^T\mathbf{x}_0 \le 0$. Then

$$(4) \qquad \int_0^{t_1} \mathbf{b}^T e^{-As}B\mathbf{u}(s)\, ds = -\mathbf{b}^T\mathbf{x}_0 \ge 0 \quad \text{for all } \mathbf{u}(\cdot) \in \mathscr{U}_m \text{ and all } \mathbf{x}_0 \in \mathscr{C}(t_1).$$

This implies (cf. the Lemma below) that $(**)$ $\mathbf{b}^T e^{-As}B \equiv \mathbf{0}$ on $[0, t_1]$. Setting $s = 0$, we have $\mathbf{b}^T B = \mathbf{0}$. Differentiating the identity $(**)$ k times and setting $s = 0$, we obtain $\mathbf{b}^T A^k B = \mathbf{0}$, $k = 0, 1, 2, \ldots$. Thus $\mathbf{b}$ is orthogonal to the columns of M, and rank $M < n$. □

Remarks. We have actually proved that for (LA):
1. rank $M < n \Rightarrow \exists$ a fixed hyperplane which contains all $\mathscr{C}(t_1)$, $t_1 > 0$;
2. rank $M = n \Leftrightarrow \mathbf{0} \in \text{Int } \mathscr{C}$ $\forall t > 0$;
3. rank $M = n \Leftrightarrow \forall \mathbf{b} \neq \mathbf{0}$, $\mathbf{b}^T e^{-At}B \not\equiv \mathbf{0}$ as a function of t.

Systems for which $\mathbf{b}^T e^{-At}B \not\equiv \mathbf{0}$ for each $\mathbf{b} \neq \mathbf{0}$, are called *proper*; for (LA) this is equivalent to rank $M = n$. This term will be defined in a later chapter for systems other than (LA), although we will not have such a nice algebraic characterization.

The reason we have separated the following lemma from the text of the proof of Theorem 3 is that it is the only place where the structure of $\mathscr{U}_m$ enters directly. The structure of $\mathscr{U}_m$ also enters indirectly in our use of the convexity of $\mathscr{C}$. This means that if we can establish the convexity of $\mathscr{C}$, and the lemma, for other classes of controls, then Theorem 3 will hold for these classes. We will discuss the extension of this lemma immediately following its proof.

Lemma. *Assume that $t_1 > 0$ and $\mathbf{b} \in R^m$ are given. If (4) holds for all $\mathbf{u}(\cdot) \in \mathscr{U}_m[0, t_1]$, then the $(1 \times m)$ row vector $\mathbf{b}^T e^{-As}B \equiv \mathbf{0}$ for $s \in [0, t_1]$.*

Proof. Let $\mathbf{v}(s) \in R^m$ be the column vector whose transpose is the row vector $\mathbf{b}^T e^{-As}B$. If $\mathbf{u}(\cdot) \in \mathscr{U}_m$, then $-\mathbf{u}(\cdot) \in \mathscr{U}_m$. But replacing $\mathbf{u}(\cdot)$ by $-\mathbf{u}(\cdot)$ in (4) changes the sign of the integral, therefore, $\int_0^{t_1} \mathbf{v}^T(s)\mathbf{u}(s)\, ds = 0$ for all $\mathbf{u}(\cdot) \in \mathscr{U}_m$. Suppose that for some $s_0 \in [0, t_1]$, $\mathbf{v}(s_0) \neq \mathbf{0} \in R^m$. Then define $\mathbf{u}(s)$ to be $\mathbf{0} \in R^m$, except for s in a small neighborhood N of s_0. In N, let $\mathbf{u}(s)$ be the constant vector $\mathbf{v}(s_0)$. By the continuity of $v(s)$, we can choose N so that $\int_0^{t_1} \mathbf{v}^T(s)\mathbf{u}(s)\, ds = \int_N \mathbf{v}^T(s)\mathbf{v}(s_0)\, ds > 0$, a contradiction. □

The above proof only required two properties of $\mathscr{U}_m$:

(i) $\mathbf{u}(\cdot) \in \mathscr{U}_m \Rightarrow -\mathbf{u}(\cdot) \in \mathscr{U}_m$;

(ii) given any $s_0 \in [0, t_1]$ and any fixed nonzero vector $\mathbf{v}(s_0) \in R^m$, there is an admissible control $\mathbf{u}_0(\cdot)$ such that $\mathbf{v}^T(s_0)\mathbf{u}_0(s)$ is positive in some neighborhood of s_0 and is zero outside this neighborhood. The classes of piecewise constant and smooth controls satisfy these requirements. Also, the unit cube Ω can be replaced by any symmetric compact subset Ψ with $\mathbf{0} \in \operatorname{Int} \Psi$.

Since controllability is a concept that is independent of the bases used in R^n, R^m respectively, it must follow that if we make the changes of variables and controls $\mathbf{y}(t) = \mathbf{Px}(t)$ and $\mathbf{v}(t) = Q\mathbf{u}(t)$ where P, Q are nonsingular constant $n \times n$, $m \times m$ matrices respectively, then the new system

$$\dot{\mathbf{y}} = PAP^{-1}\mathbf{y} + PBQ^{-1}\mathbf{v} = \tilde{A}\mathbf{y} + \tilde{B}\mathbf{v}$$

must have the same controllability properties (modulo transforming regions in R^n and R^m under the maps $P\mathbf{x}$, $Q\mathbf{u}$) as the original system (LA) $\dot{\mathbf{x}} = A\mathbf{x} + B\mathbf{u}$. This is easy to see algebraically, since

$$\operatorname{rank}\{B, AB, A^2B, \ldots, A^{n-1}B\} = \operatorname{rank} P\{B, AB, A^2B, \ldots, A^{n-1}B\}Q^{-1}.$$

We now turn to the important question of when $\mathscr{C} = R^n$ (complete controllability), that is, when can we steer *every* state $\mathbf{x}_0$ to the target $\mathbf{0} \in R^n$.

EXAMPLE 2. Consider the one-dimensional problem $\dot{x} = x + u$, $-1 \le u(t) \le 1$. Because of the constraints on $u(t)$, $\dot{x}(t) > 0$ if $x(t) > 1$ and $\dot{x}(t) < 0$ if $x(t) < -1$. Therefore, $\mathscr{C} \subset [-1, 1]$. By equation (3), $x_0 \in \mathscr{C}(t_1) \Leftrightarrow x_0 = -\int_0^{t_1} e^{-s} u(s)\, ds$, for some admissible control $u(\cdot)$. Therefore, since $|u(t)| \le 1$, $x_0 \in \mathscr{C}(t_1)$ implies that $|x_0| \le \int_0^{t_1} e^{-s}\, ds = 1 - e^{-t_1} \le 1$. If in fact $|x_0| \le 1 - e^{-t_1}$, then it is easy to solve the equation and choose $u(t)$ so that $x(t_1; x_0, u(\cdot)) = 0$. In fact, one can choose $u(t) \equiv +1 (x_0 < 0)$ or $u(t) \equiv -1 (x_0 > 0)$ on some interval $[0, t_*]$, with $u(t) \equiv 0$ on $(t_*, t_1]$. Therefore $\mathscr{C}(t_1) = \{x_0 \mid |x_0| \le 1 - e^{-t_1}\}$, and $\mathscr{C} = \{x_0 \mid |x_0| < 1\}$ (see Figure 5). For this

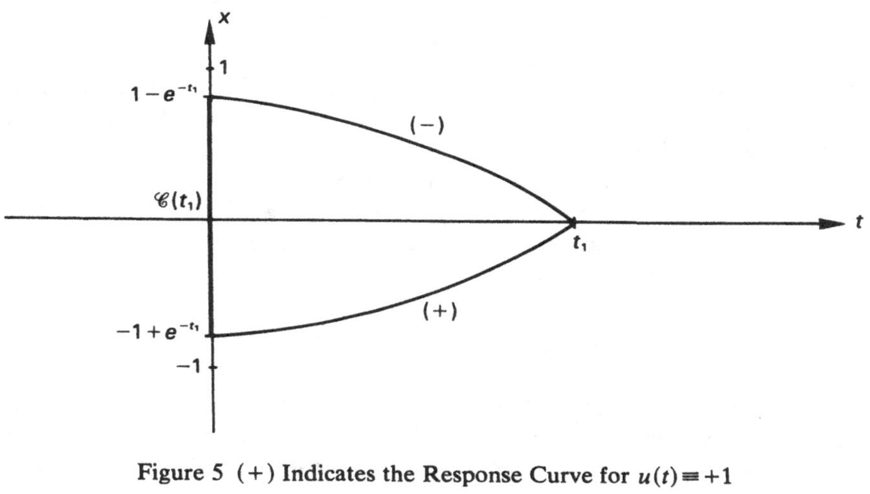

Figure 5 (+) Indicates the Response Curve for $u(t) \equiv +1$

example, then, $\mathscr{C} \neq R^n$. The problem is that the homogeneous equation $\dot{x}(t) = x(t)$ has the origin $x = 0$ as an unstable equilibrium.

Theorem 4. *Suppose that* rank $M = n$ *for the linear autonomous system* (LA) $\dot{\mathbf{x}} = A\mathbf{x} + B\mathbf{u}$. *If* Re $\lambda < 0$ *for every eigenvalue* λ *of* A, *then* $\mathscr{C} = R^n$.

Proof. We know from Theorem 3 that the origin $\mathbf{0} \in R^n$ has a neighborhood $\mathscr{B}_0 \equiv \mathscr{B}(\mathbf{0}; \delta)$ with $\mathbf{0} \in \mathscr{B}_0 \subset \mathscr{C}$. For any given $\mathbf{x}_0 \in R^n$, we first set $\mathbf{u}(t) \equiv 0$, so that the dynamics of (LA) is determined by $\dot{\mathbf{x}} = A\mathbf{x}$. By standard results on the stability of linear differential equations (cf. Mathematical Appendix), our assumption that Re $\lambda < 0$ for all eigenvalues λ of A implies that each solution of $\dot{\mathbf{x}} = A\mathbf{x}$ eventually enters and stays in $\mathscr{B}_0$. Once in $\mathscr{B}_0$, we are in $\mathscr{C}$, so we can switch to a control that will steer to the origin. □

The stability assumption in the above theorem is rather strong. For example, the rocket car has $\ddot{p}(t) = u(t)$, or in system form

$$\mathbf{x} = \begin{bmatrix} p \\ \dot{p} \end{bmatrix}, \qquad \dot{\mathbf{x}} = \begin{bmatrix} 0 & 1 \\ 0 & 0 \end{bmatrix} \mathbf{x} + u(t) \begin{bmatrix} 0 \\ 1 \end{bmatrix}, \qquad -1 \leq u(t) \leq +1,$$

so A has the eigenvalue zero with multiplicity two. Theorem 4 is therefore of no use. The following result is much deeper, in that it does not require stability.

Theorem 5. *For the linear autonomous system* (LA), *recall that* $M = [B, AB, \ldots, A^{n-1}B]$. *Then*

$$\mathscr{C} = R^n \Leftrightarrow \text{rank}\, M = n \textit{ and } \text{Re}\, \lambda \leq 0 \textit{ for each eigenvalue } \lambda \textit{ of } A.$$

Proof. Suppose that rank $M = n$ and Re $\lambda \leq 0$ for each eigenvalue λ of A. First we recall that $\mathscr{C}$ is convex (Theorem 2). If there were a state $\mathbf{w}_0 \notin \mathscr{C}$, then $\mathbf{w}_0$ could be separated from $\mathscr{C}$ by a hyperplane – that is, there would exist a fixed vector $\mathbf{b} \in R^n$ and a real number α such that

$$\langle \mathbf{b}, \mathbf{x}_0 \rangle \equiv \mathbf{b}^T \mathbf{x}_0 \leq \alpha \quad \text{for all } \mathbf{x}_0 \in \mathscr{C}.$$

We will in fact show that for any nonzero vector $\mathbf{b} \in R^n$ and any real number α there is a vector $\mathbf{x}_0 \in \mathscr{C}$ such that $\mathbf{b}^T \mathbf{x}_0 > \alpha$. This contradiction will prove half of the theorem.

Let $\mathbf{b} \neq \mathbf{0} \in R^n$ be given. We want to prove that $\mathscr{C}$ contains vectors $\mathbf{x}_0$ such that $\mathbf{b}^T \mathbf{x}_0$ is arbitrarily large. Now $\mathbf{x}_0 \in \mathscr{C}$ if and only if $\mathbf{x}_0 = -\int_0^{t_1} e^{-As} B\mathbf{u}(s)\, ds$ for some t_1 and some $\mathbf{u}(\cdot) \in \mathscr{U}_m$, so we have to show that there is a $\mathbf{u}(\cdot) \in \mathscr{U}_m$ for which

$$-\int_0^{t_1} \mathbf{b}^T e^{-As} B\mathbf{u}(s)\, ds > \alpha.$$

For convenience, we define the (column) vector $\mathbf{v}(s) \equiv (\mathbf{b}^T e^{-As} B)^T$ in R^m. By the discussion below Theorem 3, our assumption that rank $M = n$ implies that the system is proper, that is, $\mathbf{v}(s) \not\equiv \mathbf{0}$ for $s \in [0, t_1]$. We choose

$u^i(s) = -\operatorname{sgn} v^i(s)$, $i = 1, 2, \ldots, m$, with $u^i(s) = 0$ when $v^i(s) = 0$; then for any t_1, with $\mathbf{x}_0 = -\int_0^{t_1} e^{-As} B\mathbf{u}(s)\,ds$,

$$\mathbf{b}^T\mathbf{x}_0 = \int_0^{t_1} |\mathbf{v}(s)|\,ds, \quad \text{where as always,} \quad |\mathbf{v}| = \sum_{i=1}^{m} |v^i|.$$

Now some component of $\mathbf{v}(s)$, say $v^1(s)$, is nonzero somewhere, say at s_0. By continuity, $v^1(s) \neq 0$ in a neighborhood of s_0. We will show that $\int_0^\infty |v^1(s)|\,ds = +\infty$. Suppose not, and let $\phi(t) = \int_t^\infty v^1(s)\,ds$. Note that $\phi(t) \to 0$ as $t \to \infty$, and $-D\phi = v^1$, where $D = d/dt$. The matrix A satisfies its characteristic equation, $\mathcal{P}(A) = 0$, so

$$\mathcal{P}(-D)\mathbf{v}(t) \equiv \mathcal{P}(-D)\{\mathbf{b}^T e^{-At}B\} = \mathbf{b}^T e^{-At}\mathcal{P}(A)B = \mathbf{0}.$$

Therefore $\phi(t)$ is a solution of the constant coefficient equation

$$D\mathcal{P}(-D)\phi(t) = 0, \qquad \phi(t) \not\equiv 0, \quad \text{and} \quad \lim_{t\to\infty} \phi(t) = 0.$$

This means that $\phi(t)$ is a linear combination of terms of the form $p(t)\,e^{\lambda t}$, where $p(t)$ is a polynomial and λ is a root of $\lambda\mathcal{P}(-\lambda) = 0$. These roots are all non-negative (the negatives of the eigenvalues of A, with $\lambda = 0$ added), which contradicts our conclusion that $\lim_{t\to\infty} \phi(t) = 0$.

For the remaining half of the theorem, we consider the two possibilities: $\operatorname{rank} M < n$, or $\operatorname{Re} \lambda > 0$ for some λ. Suppose $\operatorname{rank} M < n$. Then by the discussion following Theorem 3, $\mathscr{C}(t)$ is contained in a fixed hyperplane for all $t > 0$, so $\mathscr{C}(t) \neq R^n$.

Finally, suppose $\operatorname{Re} \lambda_1 > 0$ for some eigenvalue λ_1 of A. We want to show $\mathscr{C} \neq R^n$. There exists a *real* nonsingular matrix Q such that $\tilde{A} = Q^{-1}AQ$ is in *real canonical form* (cf. the Mathematical Appendix)

$$\tilde{A} = \operatorname{diag}\begin{bmatrix} J_1 & & \\ & J_2 & 0 \\ & & \\ 0 & & J_k \end{bmatrix},$$

where the form of the $m_r \times m_r$ matrix J_r depends on the eigenvalue of A to which it corresponds:

$$\text{for } \lambda_k \text{ real,} \quad J_r = \begin{bmatrix} \lambda_k & & & 0 \\ 1 & \lambda_k & & \\ & 1 & \ddots & \\ 0 & & 1 & \lambda_k \end{bmatrix},$$

for $\lambda = \alpha + i\beta$ $(\beta \neq 0)$,

$$J_r = \begin{bmatrix} R & & & \\ I_2 & R & & 0 \\ & I_2 & \ddots & \\ 0 & & I_2 & R \end{bmatrix}, \qquad I_2 = \begin{bmatrix} 1 & 0 \\ 0 & 1 \end{bmatrix}, \qquad R = \begin{bmatrix} \alpha & -\beta \\ \beta & \alpha \end{bmatrix}.$$

If we define $\mathbf{x}(t)=Q\mathbf{y}(t)$ then (LA) becomes

$$\text{(RCF)} \qquad \dot{\mathbf{y}}=Q^{-1}AQ\mathbf{y}+Q^{-1}B\mathbf{u}\equiv\tilde{A}\mathbf{y}+\mathbf{w}(t), \qquad \mathbf{y}(t)\in R^n.$$

Since $|u(t)|\leq 1$, we have $|\mathbf{w}(t)|\leq K$ for some constant K. Without loss of generality, assume that λ_1 appears in J_1. If λ_1 is real, then $\dot{y}^1(t)=\lambda_1 y^1(t)+w^1(t)$ is the first equation in (RCF). If $y^1(0)>K/\lambda_1$, then $y^1(t)$ is always increasing, so $y^1(t)\not\to 0$ and $\mathbf{y}(t)\not\to\mathbf{0}$.

When $\lambda_1=\alpha+i\beta$ with $\alpha>0$ and $\beta\neq 0$ the first two equations in (RCF) are

$$\dot{y}^1=\alpha y^1-\beta y^2+w^1(t),$$
$$\dot{y}^2=\beta y^1+\alpha y^2+w^2(t)$$

(superscripts denote components). Multiplying the first equation by $y^1(\cdot)$, the second by $y^2(\cdot)$, and adding, we get (recall that $\|\cdot\|$ stands for the euclidean norm)

$$\frac{1}{2}\frac{d}{dt}\|\mathbf{z}\|^2=\alpha\|\mathbf{z}\|^2+\mathbf{z}^T\mathbf{w}(t), \qquad \mathbf{z}=\begin{bmatrix} y^1 \\ y^2 \end{bmatrix}, \qquad \mathbf{w}=\begin{bmatrix} w^1 \\ w^2 \end{bmatrix}.$$

The Cauchy–Schwarz inequality implies that $|\mathbf{z}^T\mathbf{w}(t)|\leq\|\mathbf{z}\|\,\|\mathbf{w}(t)\|$, and this combined with the last equation above implies $d/dt\|\mathbf{z}(t)\|\geq\alpha\|\mathbf{z}(t)\|-K$. Therefore, if $\|\mathbf{z}(0)\|>(K/\alpha)$, it follows that $\|\mathbf{z}(t)\|$ is always increasing and so again $\mathbf{y}(t)\not\to 0$ as $t\to\infty$.

Finally, since $\mathbf{x}(t)=Q\mathbf{y}(t)$, in both of the above cases we have $\mathbf{x}(t)\not\to 0$ for the corresponding solution $\mathbf{x}(t)$ of (LA). □

EXAMPLE 3 (Controllability of the Rocket Car). For the rocket car, we have

$$A=\begin{bmatrix} 0 & 1 \\ 0 & 0 \end{bmatrix}, \qquad B=\begin{bmatrix} 0 \\ 1 \end{bmatrix}, \qquad M=[B, AB]=\begin{bmatrix} 0 & 1 \\ 1 & 0 \end{bmatrix},$$

so rank $M=2$, and the system is proper, by the remarks following Theorem 3. The only eigenvalue of A is $\lambda=0$. By Theorem 5, $\mathscr{C}=R^2$, that is, *for the rocket car, any initial state can be steered to* $\mathbf{0}\in R^2$.

It is interesting that Theorem 5 becomes considerably simpler for the case of *unrestricted controls* ($\mathbf{u}(t)\in R^m$):

Theorem 5′. *For (LA) with unrestricted controls,* $\mathscr{C}=\mathbf{R}^n \Leftrightarrow \operatorname{rank} M=n$.

The reader is asked to prove this in Exercise 6 at the end of this chapter.

We now prove a result on the generic nature of controllable systems, i.e., we show that "practically all" systems are controllable. To make this idea precise, we define the *distance between two systems*:

$$\text{(LA 1)} \qquad \dot{\mathbf{x}}=A_1\mathbf{x}+B_1\mathbf{u}$$

and

(LA 2) $$\dot{\mathbf{x}} = A_2\mathbf{x} + B_2\mathbf{u},$$

to be $|A_1 - A_2| + |B_1 - B_2|$. (Recall that $|A| = \sum |a_{ij}|$.) This makes the set of all systems $\{\dot{\mathbf{x}} = A\mathbf{x} + B\mathbf{u} | A$ an $n \times n$ constant matrix, B an $n \times m$ constant matrix$\}$ into a metric space. The two systems (LA 1), (LA 2) are "close" in this sense if each entry in A_1, B_1 is "close" to the corresponding entry in A_2, B_2. We distinguish between our usual *restricted controls* ($\mathbf{u}(t) \in \Omega$) and the case of *unrestricted controls* ($\mathbf{u}(t) \in R^m$).

Theorem 6. *With unrestricted controls, the set of all completely controllable (i.e., $\mathscr{C} = R^n$) linear autonomous systems is open and dense in the metric space of all linear autonomous systems. With restricted controls, the set of systems for which $\mathscr{C} \supset \mathscr{N}(\mathbf{0})$ is open and dense; here $\mathscr{N}(\mathbf{0})$ denotes a neighborhood of the origin which may depend on $\mathbf{u}(\cdot)$.*

Remarks. This theorem roughly states, first, that if a system is controllable, then so are *all* nearby systems (open-ness). This is important, since any model of a real world system is subject to imprecisions and perturbations. Second, the theorem roughly states that if a system is not controllable, then there are systems arbitrarily near it that are controllable (denseness), so lack of controllability is something of an accident.

Proof. We treat only the case of unrestricted controls, using the criterion from Theorem 5′ (for restricted controls, we would use the same proof, with the criterion from Theorem 3). By Theorem 5′, a given system $\dot{\mathbf{x}} = A\mathbf{x} + B\mathbf{u}$ is completely controllable with unrestricted controls if and only if

$$\operatorname{rank} M = \operatorname{rank}\{B, AB, A^2B, \ldots, A^{n-1}B\} = n.$$

This rank condition implies the determinantal condition (cf. the Mathematical Appendix):

There exists N, an $n \times n$ submatrix of M, such that $\det N \neq 0$.

If $(\tilde{A}, \tilde{B})$ is a fixed system close to (A, B), then each matrix $\tilde{Q}$ from the *finite* collection of $n \times n$ submatrices of $\tilde{M} = \{\tilde{B}, \tilde{A}\tilde{B}, \ldots, \tilde{A}^{n-1}\tilde{B}\}$ is close to the corresponding submatrix Q of M. Also, if $\det Q \neq 0$, then $\det \tilde{Q} \neq 0$ for $|Q - \tilde{Q}|$ sufficiently small. Thus for $(\tilde{A}, \tilde{B})$ sufficiently close to (A, B) we will have $\operatorname{rank} \tilde{M} = n$, and so the controllable systems form an open set.

Now suppose that the system $\dot{\mathbf{x}} = A_0\mathbf{x} + B_0\mathbf{u}$ is not completely controllable, that is $\operatorname{rank} M < n$. We need to find matrices $(\tilde{A}, \tilde{B})$ close to (A_0, B_0) such that $\det \tilde{N} \neq 0$ for some $n \times n$ submatrix of $\tilde{M} = \{\tilde{B}, \tilde{A}\tilde{B}, \ldots, \tilde{A}^{n-1}\tilde{B}\}$. Det $\tilde{N}$, for an $n \times n$ submatrix $\tilde{N}$ of $\tilde{M}$, can be thought of as a polynomial $\mathscr{P}(y^1, y^2, \ldots, y^k)$, $k = n^2 + mn$, in the entries of $\tilde{A}$ and $\tilde{B}$. In the present situation $\mathscr{P}$ vanishes when we use the entries of A_0, B_0 for $y^1, \ldots, y^k$.

Since we only need $\det \tilde{N} \neq 0$ for *some* submatrix, we now need only show:

> *Given a nontrivial polynomial $\mathcal{P}(y^1, \ldots, y^k)$ in k variables which vanishes at $\mathbf{y}_0 = (y_0^1, y_0^2, \ldots, y_0^k)$, there are vectors $\boldsymbol{\xi} = (\xi^1, \xi^2, \ldots, \xi^k)$ arbitrarily close to $\mathbf{y}_0$ such that $\mathcal{P}(\boldsymbol{\xi}) \neq 0$.*

But this is equivalent to the obvious fact that a nontrivial polynomial in k variables cannot vanish identically in any k-dimensional ball. (If it did, we could take enough partial derivatives to conclude all coefficients are zero.) □

3. Controllability for Nonlinear Autonomous Systems

We will now investigate the nonlinear autonomous system

$$\text{(NLA)} \qquad \dot{\mathbf{x}} = \mathbf{f}(\mathbf{x}, \mathbf{u}), \qquad \mathbf{u}(\cdot) \in \mathcal{U}_m, \qquad \mathcal{T}(t) \equiv 0,$$

with $\mathbf{f}(\mathbf{x}, \mathbf{u})$ continuously differentiable in $\mathbf{x}$, $\mathbf{u}$, and $\mathbf{f}(\mathbf{0}, \mathbf{0}) = \mathbf{0} \in R^n$. Our assumption that $\mathbf{f}(\mathbf{0}, \mathbf{0}) = \mathbf{0}$ is not unreasonable, since it just means that when we arrive at the target $\mathbf{0} \in R^n$, we can stay there with $\mathbf{u}(t) \equiv \mathbf{0} \in R^m$. For $\mathbf{x}(t)$ near $\mathbf{0} \in R^n$ we expect to be able to use small controls, so $\mathbf{u}(t)$ should be near $\mathbf{0} \in R^m$. We therefore expand $\mathbf{f}(\mathbf{x}, \mathbf{u})$ about $\mathbf{x} = \mathbf{0}$, $\mathbf{u} = \mathbf{0}$:

$$f(\mathbf{x}, \mathbf{u}) = \mathbf{f}_{\mathbf{x}}(\mathbf{0}, \mathbf{0})\mathbf{x} + \mathbf{f}_{\mathbf{u}}(\mathbf{0}, \mathbf{0})\mathbf{u} + o(|\mathbf{x}| + |\mathbf{u}|),$$

where $\mathbf{f}_{\mathbf{x}}(\cdot, \cdot)$, $\mathbf{f}_{\mathbf{u}}(\cdot, \cdot)$ are the $n \times n$, $n \times m$ Jacobian matrices $[\partial f^i / \partial x^j]$, $[\partial f^i / \partial u^j]$, respectively. We expect the controllability of (NLA) near $\mathbf{0} \in R^n$ to be determined by the controllability of the (autonomous) *linearization*:

$$(5) \qquad \dot{\mathbf{x}} = \mathbf{f}_{\mathbf{x}}(\mathbf{0}, \mathbf{0})\mathbf{x} + \mathbf{f}_{\mathbf{u}}(\mathbf{0}, \mathbf{0})\mathbf{u} = A_f \mathbf{x} + B_f \mathbf{u}.$$

Recall that the *controllability matrix* for (5) is

$$M_f = \{B_f, A_f B_f, A_f^2 B_f, \ldots, A_f^{n-1} B_f\}.$$

Theorem 7. *If* $\operatorname{rank} M_f = n$, *then* $\mathbf{0} \in \operatorname{Int} \mathscr{C}$ *for (NLA).*

Before we prove this theorem, some comments are in order. The converse is false, as Example 4 (following the proof) will show. In the proof, we will use the concept of *time-reversal* as discussed in the introduction. If $\mathbf{x}(t)$ solves (NLA) with $\mathbf{x}(0) = \mathbf{x}_0$, $\mathbf{x}(t_1) = \mathbf{x}_1$, then $\mathbf{z}(t) = \mathbf{x}(t_1 - t)$ solves

$$(6) \qquad \dot{\mathbf{z}} = -\mathbf{f}(\mathbf{z}, \tilde{\mathbf{u}}), \qquad \mathbf{z}(0) = \mathbf{x}_1, \qquad \mathbf{z}(t_1) = \mathbf{x}_0, \qquad \tilde{\mathbf{u}}(t) = \mathbf{u}(t_1 - t).$$

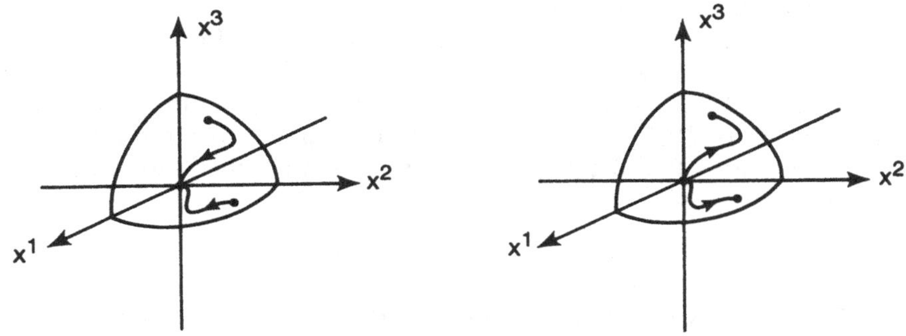

Two Responses under (NLA) The Same Responses under (6)

Figure 6

Thus the response curves are identical for (NLA) and (6), *but are traversed in opposite directions* (Figure 6). It follows that $\mathbf{x}_0$ can be steered to $\mathbf{0}$ under (NLA) if and only if $\mathbf{x}_0$ is in the reachable set from $\mathbf{0}$ under (6). To say that any point in a neighborhood $\mathcal{B}(\mathbf{0}, \delta)$ of $\mathbf{0} \in R^n$ can be steered to $\mathbf{0}$ for (NLA) is equivalent to saying: "the fixed initial value $\mathbf{0} \in R^n$ can be successfully steered under (6) to any point in $\mathcal{B}(\mathbf{0}; \delta)$." We used $K(t_1; \mathbf{x}^*)$ to stand for the reachable set at time t_1 from the initial state $\mathbf{x}^*$ under (NLA), using all possible controls from $\mathcal{U}_m$. We define $K^-(t_1)$ to stand for the reachable set at time t_1 starting from $\mathbf{0} \in R^n$, under (6) (we omit reference to $\mathbf{x}^*$ since $\mathbf{x}^* = \mathbf{0}$ always).

We also need the following technical lemma, which states that if we can cover a neighborhood of $\mathbf{x}_0 = \mathbf{0}$ under the time-reversed system (6), then we can in fact specify our time of arrival at certain state vectors.

Lemma. *Suppose that the time-reversed linear autonomous system*

$$\dot{\mathbf{x}} = -A\mathbf{x} - B\mathbf{u}, \qquad \mathbf{x}(0) = \mathbf{0}, \tag{7}$$

can steer $\mathbf{0} \in R^n$ to any point in a ball $\mathcal{B}(\mathbf{0}; \delta)$. Then given a set of scalar multiples of the canonical basis vectors

$$S = \{\alpha\mathbf{e}_1, \alpha\mathbf{e}_2, \ldots, \alpha\mathbf{e}_n\} \quad \textit{with } \alpha > 0 \textit{ sufficiently small,}$$

we can find controls $\mathbf{u}_1(\cdot), \ldots, \mathbf{u}_n(\cdot)$ in $\mathcal{U}_m$ such that under $\mathbf{u}_j(\cdot)$ we hit $\alpha\mathbf{e}_j$ at $t = 1$:

$$\mathbf{x}(1; \mathbf{0}, \mathbf{u}_j(\cdot)) = \alpha\mathbf{e}_j, \qquad j = 1, \ldots, n.$$

The proof is outlined in Exercise 1 at the end of this chapter.

Proof of Theorem 7. We will prove that $\mathbf{0} \in R^n$ can be steered to any point in some open ball $\mathcal{B}_0 \equiv \mathcal{B}(\mathbf{0}; \delta) \subset R^n$ under the time-reversed system (6); in fact we will show that $K^-(1) \supset \mathcal{B}_0$ for some $\delta > 0$, so the time of arrival at each point of $\mathcal{B}_0$ can be specified as $t = 1$.

We cannot be sure that solutions of (6) with $\mathbf{x}(0)=\mathbf{0}$ extend to $0 \le t \le 1$. However, because $\mathbf{f}(\mathbf{x}, \mathbf{u})$ is differentiable, it follows that for $\sup_{[0,1]} |\mathbf{u}(t)|$ sufficiently small the response $\mathbf{x}(t; \mathbf{0}, \mathbf{u}(\cdot))$ will exist on $[0, 1]$. To see this, note that $\mathbf{x}[t] \equiv \mathbf{0} \in R^n$ is the unique response to $\mathbf{u}(t) \equiv \mathbf{0}$, and this response certainly exists on $[0, 2]$. Therefore, since the response to $\mathbf{u}(t) \equiv \mathbf{0}$ exists on $[0, 2]$, it follows that there exists $\varepsilon > 0$ such that (cf. the Mathematical Appendix):

$$(*) \qquad \sup_{[0,1]} |u(t)| < \varepsilon \Rightarrow \mathbf{x}(t; \mathbf{0}, \mathbf{u}(\cdot)) \quad \text{exists on } [0, 1].$$

By Theorem 3, $\mathbf{0} \in \operatorname{Int} \mathscr{C}$ for the linear system (5); therefore under the *linear time-reversed* system (7), the origin $\mathbf{0} \in R^n$ can be steered to any point in some fixed neighborhood of $\mathbf{0} \in R^n$. Then, by the lemma above, for some $\alpha > 0$ there are controls $\mathbf{u}_1(\cdot), \ldots, \mathbf{u}_n(\cdot)$ such that $\mathbf{x}_k[1] \equiv \mathbf{x}(1; \mathbf{0}, \mathbf{u}_k(\cdot)) = \alpha \mathbf{e}_k$, $(k = 1, \ldots, n)$ under the linear time-reversed system (7). Unfortunately, $\sup_{[0,1]} |\mathbf{u}(t)|$ may exceed the extendability restriction in (*) above. To deal with this problem, we define a parametrized family of controls

$$\mathbf{u}(t, \mathbf{c}) = c^1 \mathbf{u}_1(t) + \cdots + c^n \mathbf{u}_n(t),$$

where $\mathbf{c} \in R^n$ is restricted by $\|\mathbf{c}\| \equiv (\sum_1^n [c^i]^2)^{1/2} < \varepsilon/n$. Then each c^i satisfies $|c^i| < \varepsilon/n$, and since $|\mathbf{u}_k(t)| \le 1$, we have $|\mathbf{u}(t, \mathbf{c})| < \varepsilon$, $0 \le t \le 1$.

We now consider the family of responses to the controls $\mathbf{u}(t; \mathbf{c})$:

$$\mathbf{z}(t; \mathbf{0}, \mathbf{u}(t; c))$$

under the *nonlinear time-reversed* system (6), $0 \le t \le 1$. We define a function $\mathbf{g}(\mathbf{c})$:

$$\mathbf{g}(\cdot)\colon \mathscr{B}_0 \equiv \mathscr{B}(0; \varepsilon/n) \to R^n \quad \text{by } \mathbf{g}(\mathbf{c}) = \mathbf{z}(1, \mathbf{c}),$$

i.e., the point in R^n reached at $t = 1$ by the response to $\mathbf{u}(\cdot, \mathbf{c})$ under (6). Clearly $\mathbf{g}(\mathbf{0}) = \mathbf{z}(1, \mathbf{0}) = \mathbf{0}$ (the response under (6) to $\mathbf{u}(t) \equiv \mathbf{0}$ is $\mathbf{x}[t] \equiv \mathbf{0}$, since $\mathbf{x}_0 = \mathbf{0}$). If we can show that the Jacobian matrix $J = \mathbf{g}_\mathbf{c}(\mathbf{0})$ is nonsingular, then $\mathbf{g}(\cdot)$ will map open sets into open sets, in particular, the image of $\mathscr{B}_0$ will contain a neighborhood $\mathscr{B}_1$ of $\mathbf{g}(\mathbf{0}) = \mathbf{0} \in R^n$. This will imply that the set of points reached at $t = 1$ by the responses under (6) to the family of controls $\{\mathbf{u}(t, \mathbf{c}) | \mathbf{c} \in \mathscr{B}_0\}$ contains a neighborhood of $\mathbf{0}$ (see Figure 7).

To complete the proof then, we need to show that J is nonsingular. We have by definition

$$(**) \qquad \dot{\mathbf{z}}(t, \mathbf{c}) = -\mathbf{f}(\mathbf{z}, \mathbf{u}(t, \mathbf{c})), \qquad \mathbf{z}(0, \mathbf{c}) = \mathbf{0}.$$

Let the $n \times n$ matrix $N(t)$ be defined by

$$N(t) = \mathbf{z}_\mathbf{c}(t, \mathbf{c})|_{\mathbf{c}=\mathbf{0}}, \quad \text{so } J = N(1).$$

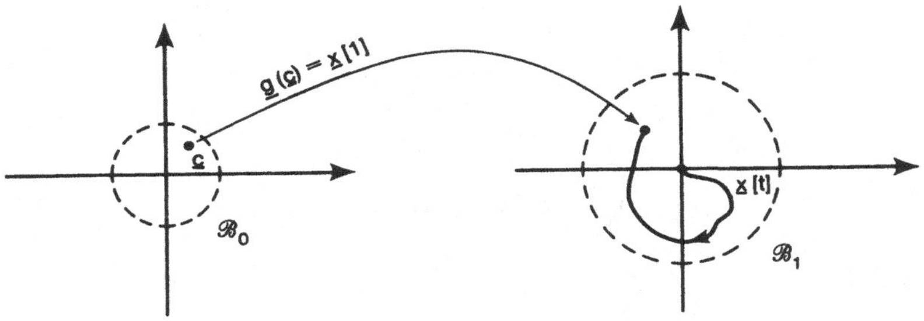

Figure 7 Each Point $\mathbf{c} \in \mathscr{B}_0$ Defines a Control $\mathbf{u}(t, \mathbf{c})$. This Control in Turn Generates a Nonlinear Response $\mathbf{z}[t]$, and $\mathbf{g}(\mathbf{c}) = \mathbf{z}[1]$

Then $N(0) = 0$, and we can take partial derivatives with respect to the c^i in (**) to see that N satisfies the linear system

$$\dot{N}(t) = -\mathbf{f_x}(\mathbf{0}, \mathbf{0})N(t) - \mathbf{f_u}(\mathbf{0}, \mathbf{0})\mathbf{u_c}(t, \mathbf{c})\big|_{\mathbf{c}=\mathbf{0}},$$

i.e.,

$$\dot{N}(t) = -\mathbf{f_x}(\mathbf{0}, \mathbf{0})N(t) - \mathbf{f_u}(\mathbf{0}, \mathbf{0})[\mathbf{u}_1(t), \mathbf{u}_2(t), \ldots, \mathbf{u}_n(t)].$$

Then each column $\mathbf{n}_j(t)$ of $N(t)$ satisfies (7):

$$\mathbf{n}_j(t) = -A_f\mathbf{n}_j(t) - B_f\mathbf{u}_j(t), \qquad \mathbf{n}_j(0) = \mathbf{0},$$

and therefore coincides with the response $\mathbf{z}(t; \mathbf{0}, \mathbf{u}_j(\cdot))$ to $\mathbf{u}_j(\cdot)$ under the linear time-reversed system (7). But for this system $\mathbf{u}_j(\cdot)$ steers $\mathbf{0}$ to $\alpha\mathbf{e}_j$ at $t = 1$, therefore $\mathbf{n}_j(1) = \alpha\mathbf{e}_j$, $j = 1, \ldots, n$. Thus $N(1) = \alpha I$ and so $\mathbf{z_c}(1, \mathbf{c})\big|_{\mathbf{c}=\mathbf{0}} = N(1)$ is nonsingular. □

Example 4. As promised, we show that $\mathbf{0} \in \operatorname{Int} \mathscr{C}$ for (6) does not imply rank $M_f = n$. Consider the following two-dimensional system with scalar control: $\dot{p} = -p + u$, $\dot{q} = -q - [p(t)]^3$, or in vector form

$$\mathbf{x} = \begin{bmatrix} p \\ q \end{bmatrix}, \qquad \dot{\mathbf{x}} = \begin{bmatrix} -1 & 0 \\ 0 & -1 \end{bmatrix}\mathbf{x} + u(t)\begin{bmatrix} 1 \\ 0 \end{bmatrix} - \begin{bmatrix} 0 \\ [p(t)]^3 \end{bmatrix}.$$

First we show that $\mathbf{0} \in \operatorname{Int} \mathscr{C}$. We restrict our attention to the domain $|p(t)| \le 1$, $|q(t)| \le 1$. If we set $u(t) \equiv -1$, responses within this domain move towards the equilibrium point $(-1, 1)$ as sketched in Figure 8. For $\mathbf{x}_0^T = (p_0, q_0) = (1, 1)$, the response to $u(t) \equiv -1$ successfully reaches the target $(0, 0)$, and the part of this trajectory from $(1, 1)$ to $(0, 0)$ we denote by Γ_-.

For $u(t) \equiv +1$, a dual analysis applies, as sketched in Figure 9. In this case we have a single successful response, Γ_+, from $(-1, -1)$ to $(0, 0)$.

We will show that we can steer any initial state in some small disc about the origin so as to hit one of the trajectories Γ_+ or Γ_-. Once we hit Γ_+ or Γ_-, we can steer to the target $\mathbf{0}$ by using $u(t) \equiv +1$ or $u(t) \equiv -1$ respectively.

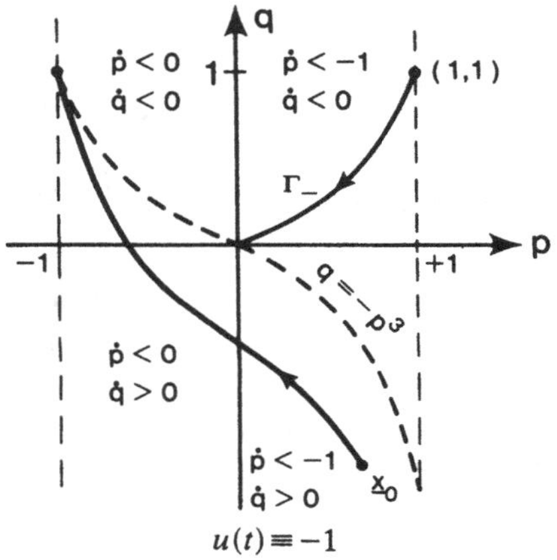

$u(t) \equiv -1$

Figure 8

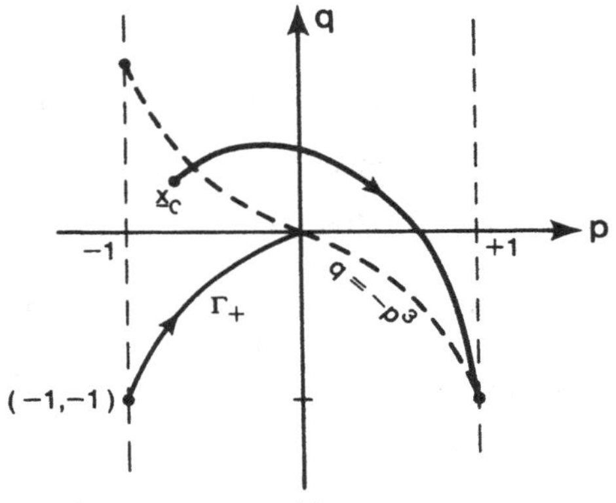

$u(t) \equiv +1$

Figure 9

As a first step, we claim that if $\mathbf{x}_0$ lies above $\Gamma_+ \cup \Gamma_-$ with $\|\mathbf{x}_0\| < 1$ and $p_0 < 0$, then we can choose $u(t) \equiv +1$ and have $p(t)$ positive before $\|\mathbf{x}(t)\|$ exceeds 1. Thus responses with sufficiently small initial states will remain in the disc D of radius 1 at least until $p(t)$ becomes positive. To see this, we write $\dot{p} = -p + 1$, $\dot{q} = -q - [p]^3$, and set $r(t) = \{[p(t)]^2 + [q(t)]^2\}^{1/2}$. Then

$$r\dot{r} = p\dot{p} + q\dot{q} = -[r(t)]^2 + p - q[p]^3.$$

As long as $p(t) \le 0$, we have $r\dot{r} \le -[r]^2 + 1/2\{[q]^2 + [p]^6\}$ (using the inequality $2ab \le [a]^2 + [b]^2$), thus $\dot{r}r \le -[r]^2 + 1/2[r]^2 + 1/2[r]^6$. So as long as $p(t) \le 0$ and $r(t) < 1$, we have $\dot{p}(t) \ge 1$ and $\dot{r}(t) \le 0$. Therefore, $p(t)$ will be positive before $t_* = 2|p_0|$, yet for $0 \le t \le t_*$

$$r(t) \le r(0) < 1.$$

Therefore, if $\mathbf{x}_0$ is sufficiently close to $(0, 0)$ and lies above $\Gamma_+ \cup \Gamma_-$, we can set $u(t) \equiv +1$ and still remain in D, until the response reaches a state (p_1, q_1) with $p_1 > 0$. We then set $u(t) \equiv p_1$, so $\dot{p}(t) = 0$, $\dot{q}(t) = -q(t) - [p_1]^3$. The trajectory therefore will move straight down until it intersects Γ_-, at which instant we switch to $u(t) \equiv -1$; the response will then move to $(0, 0)$ (Figure 10). A dual argument works for $\mathbf{x}_0$ below $\Gamma_+ \cup \Gamma_-$.

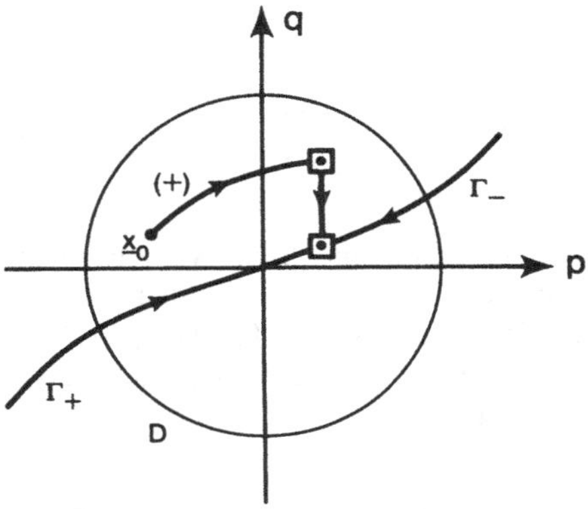

Figure 10

Thus the controllable set $\mathscr{C}$ for (NLA) contains a neighborhood of $(0, 0)$. However, the linearized problem is $\dot{p} = -p + u$, $\dot{q} = -q$, so

$$A_f = \begin{bmatrix} -1 & 0 \\ 0 & -1 \end{bmatrix}, \qquad B_f = \begin{bmatrix} 1 \\ 0 \end{bmatrix}, \qquad M_f \equiv [B_f, A_f B_f] = \begin{bmatrix} 1 & -1 \\ 0 & 0 \end{bmatrix},$$

and rank $M_f = 1$.

Theorem 8. *For the nonlinear system (NLA), suppose* rank $M_f = n$. *If the solution* $\mathbf{x}(t) \equiv \mathbf{0}$ *of the free system* $\dot{\mathbf{x}} = \mathbf{f}(\mathbf{x}, \mathbf{0})$ *is* globally asymptotically stable, *then* $\mathscr{C} = R^n$ *for (NLA).*

Proof. The previous theorem guarantees the existence of $\delta > 0$ such that $\mathscr{B}(\mathbf{0}; \delta) \subset \mathscr{C}$ for (NLA). Global asymptotic stability of the free $(\mathbf{u}(t) \equiv \mathbf{0})$ equation means that $\lim_{t \to \infty} \mathbf{x}(t; \mathbf{x}_0, \mathbf{0}) = \mathbf{0}$ for any $\mathbf{x}_0 \in R^n$, thus each solution can be "steered" (using $\mathbf{u}(t) \equiv \mathbf{0}$) to inside $\mathscr{B}(\mathbf{0}; \delta)$ *within a finite time.* Because the system is autonomous, we can then use the fact that $\mathscr{B}(\mathbf{0}; \delta) \subset \mathscr{C}$ to steer to $\mathbf{0}$ in a finite time. □

EXAMPLE 5 (Application to a Nonlinear Damped Spring). Consider the scalar equation $\ddot{p}(t) + g(p, \dot{p})\dot{p}(t) + h(p) = u(t)$, $-1 \le u(t) \le 1$, with $g(\cdot, \cdot)$ and $h(\cdot)$ continuously differentiable (so solutions to initial value problems are unique). This is equivalent to the following two-dimensional system with scalar control:

$$\text{(8)} \qquad \mathbf{x}(t)=\begin{bmatrix} p(t) \\ q(t) \end{bmatrix}, \qquad \dot{\mathbf{x}}=\begin{bmatrix} 0 & 1 \\ 0 & -g(p,q) \end{bmatrix}\mathbf{x}+\begin{bmatrix} 0 \\ -h(p)+u(t) \end{bmatrix}.$$

The linearized system is

$$\text{(9)} \qquad \dot{\mathbf{x}}=\begin{bmatrix} 0 & 1 \\ -h'(0) & -g(0,0) \end{bmatrix}\mathbf{x}+\begin{bmatrix} 0 \\ 1 \end{bmatrix}u,$$

and

$$M_f=\begin{bmatrix} 0 & 1 \\ 1 & -g(0,0) \end{bmatrix}, \qquad \operatorname{rank} M_f=2.$$

Therefore, under any conditions which will imply that (8) with $\mathbf{u}(t)\equiv 0$ is globally asymptotically stable (g.a.s.) it will follow that $\mathscr{C}=R^2$ for (8). For example, if $g(p,q)>0$ for all (p,q), $ph(p)>0$ for $p\neq 0$, and $\lim_{|p|\to\infty}\int_0^p h(s)\,ds=+\infty$, then the system is g.a.s. We sketch a proof, using a Liapunov function. Define

$$V(p,q)=\frac{q^2}{2}+H(p), \quad \text{where } H(p)=\int_0^p h(s)\,ds.$$

Then

$$V(p,q)>0 \quad \text{for all } (p,q)\neq(0,0), \qquad \lim_{|x|\to\infty} V(p,q)=+\infty,$$

and along free $(u(t)\equiv 0)$ solutions of (8),

$$\dot{V}(p,q)\equiv\frac{d}{dt}V(p(t),q(t))=q\dot{q}+h(p)\dot{p}=-q^2 g(p,q)<0.$$

Then standard results from stability theory (Hahn [1967], Theorem 26.2, p. 108) imply that (8) is g.a.s.

4. Special Controls

In the preceding sections we established basic results for both linear and nonlinear systems when $\mathbf{u}(\cdot)\in\mathscr{U}_m$, that is, when $\mathbf{u}(\cdot)$ is measurable on some interval $[0,t_1]$ with range $\mathbf{u}(t)\in\Omega$, the unit cube in R^m. Such functions can be quite pathological from the practical point of view, and therefore difficult (or impossible) to synthesize. It is an important problem, therefore, to establish controllability results when the set of admissible controls is one of the classes $\mathscr{U}_{\text{PC}}$, $\mathscr{U}_{\varepsilon}$, $\mathscr{U}_{\text{BB}}$, $\mathscr{U}_{\text{BBPC}}$ introduced in Chapter I. We define the corresponding controllable sets $\mathscr{C}_{\text{PC}}$, $\mathscr{C}_{\varepsilon}$, $\mathscr{C}_{\text{BB}}$, $\mathscr{C}_{\text{BBPC}}$.

For convenience, we briefly summarize our principal results for (LA) and (NLA) when $\mathbf{u}(\cdot) \in \mathcal{U}_m$. Recall that (LA) is *proper* if $\mathbf{b}^T e^{-At} B \not\equiv \mathbf{0}$ for any $\mathbf{b} \neq \mathbf{0}$; this is equivalent to rank $[B, AB, \ldots, A^{n-1}B] = n$.

Theorem 1. *For (NLA), $\mathscr{C}$ is arcwise connected, and $\mathscr{C}$ is open* $\Leftrightarrow \mathbf{0} \in \operatorname{Int} \mathscr{C}$.

Theorem 2. *For (LA), $\mathscr{C}$ is symmetric and convex.*

Theorem 3. *For (LA),*

(i) rank $M < n \Leftrightarrow \mathscr{C}(t_1)$ *is contained in a hyperplane.*
(ii) rank $M = n \Leftrightarrow \mathbf{0} \in \operatorname{Int} \mathscr{C}$.

Theorem 5. *(LA) is proper and* Re $\lambda \leq 0$ *for all eigenvalues of* $A \Leftrightarrow \mathscr{C} = R^n$.

Theorem 7. *For (NLA),* rank $M_f = n \Rightarrow \mathbf{0} \in \operatorname{Int} \mathscr{C}$.

Theorem 8. *For (NLA), if the free* ($\mathbf{u}(t) \equiv \mathbf{0}$) *system is globally asymptotically stable, and* rank $M_f = n$ *then* $\mathscr{C} = R^n$.

We can then summarize what is known about the extension of each theorem to the cases when our control classes are $\mathcal{U}_{\text{PC}}$, $\mathcal{U}_\varepsilon$, $\mathcal{U}_{\text{BB}}$, $\mathcal{U}_{\text{BBPC}}$.

Theorems 1, 2, 3(i), 5, and 8 are valid for *all* of the classes $\mathscr{C}_{\text{PC}}$, $\mathscr{C}_\varepsilon$, $\mathscr{C}_{\text{BB}}$, $\mathscr{C}_{\text{BBPC}}$. The extensions are direct, paralleling the given proofs, for all but Theorem 2. Theorem 2 is a special case of the theorem from Lee and Markus which was stated just after the proof of Theorem 2.

Theorem 3(ii) holds for $\mathscr{C}_\varepsilon$ and $\mathscr{C}_{\text{PC}}$. Our proof of Theorem 3(ii) fails to extend to the class $\mathcal{U}_{\text{BB}}$, because we needed to use a control which takes the value zero.

In Example 3, we showed that rank $M = n$, and Re $\lambda \leq 0$ for all eigenvalues of A, for the rocket car. Therefore, $\mathscr{C}_\varepsilon = \mathscr{C}_{\text{PC}} = R^2$ for the rocket car, by Theorems 3 and 5. This means, for example, that for the rocket car we can steer to $\mathbf{0}$ from any initial state, using a control which changes values smoothly and as slowly as we wish.

Theorem 7 holds with any class of controls for which we can prove the associated lemma. The proof of this lemma is outlined for $\mathcal{U}_m$ in Exercise 1. Theorem 7 fails for $\mathcal{U}_{\text{BB}}$, as the following example shows.

Example 6 (Linearization Does Not Give Full Information About $\mathscr{C}_{\text{BB}}$). Consider the scalar ($n = m = 1$) equation $\dot{x}(t) = u(t) + [u(t)]^2$, with $-1 \leq u(t) \leq 1$. Then

$$A_f = f_x(0,0) = 0, \qquad B_f = f_u(0,0) = 1, \qquad M_f = 1$$

so rank $M_f = 1 = n$. Thus 0 lies in the interior of $\mathscr{C}_{\text{PC}}$, $\mathscr{C}_\varepsilon$, and $\mathscr{C}$ (Theorem 7). But $u(\cdot) \in \mathcal{U}_{\text{BB}}$ implies that $u(t) + [u(t)]^2$ is either 0 or 2, so $\dot{x}(t) \geq 0$.

Thus $0 \notin \text{Int } \mathscr{C}_{BB}$ since no point $x_0 > 0$ can be steered to 0.

There is a close relationship between $\mathscr{C}_{PC}$ and $\mathscr{C}$, as the following results show. Note that the system (NLA) contains (LA) as a special case.

Theorem 9. *For (NLA), if* $\mathbf{0} \in \text{Int } \mathscr{C}_{PC}$ *then* $\mathscr{C}_{PC} = \mathscr{C}$. *In other words, if* $\mathbf{0} \in \text{Int } \mathscr{C}_{PC}$, *then anything you can accomplish with measurable controls can, in fact, be accomplished with piecewise constant controls.*

Proof. Let $\mathbf{x}_0 \in \mathscr{C}$ and choose $\mathbf{u}(\cdot) \in \mathscr{U}_m$ steering $\mathbf{x}_0$ to $\mathbf{0}$ on $[0, t_1]$. We must show that there is a $\mathbf{u}_0(\cdot) \in \mathscr{U}_{PC}$ that steers $\mathbf{x}_0$ to $\mathbf{0}$. Choose $\delta > 0$ so small that the ball $\mathscr{B}_0 \equiv \mathscr{B}(\mathbf{0}; \delta) \subset \mathscr{C}_{PC}$ in R^n. Define (the shaded region in Figure 11) the δ-tube

$$\mathscr{N}_\delta = \{\mathbf{x} \in R^n \mid \inf_{[0,t_1]} |\mathbf{x} - \mathbf{x}(t; \mathbf{x}_0, \mathbf{u}(\cdot))| \le \delta\},$$

and $\mathscr{N} = \mathscr{N}_\delta \times \Omega \subset R^{n+m}$. The idea is to approximate $\mathbf{u}(\cdot)$ by a $\mathbf{u}_0(\cdot) \in \mathscr{U}_{PC}$ so closely that the resulting response $\mathbf{x}(t; \mathbf{x}_0, \mathbf{u}_0(\cdot))$ stays in $\mathscr{N}_\delta$ and is pulled into the ball $\mathscr{B}_0 \subset \mathscr{C}_{PC}$.

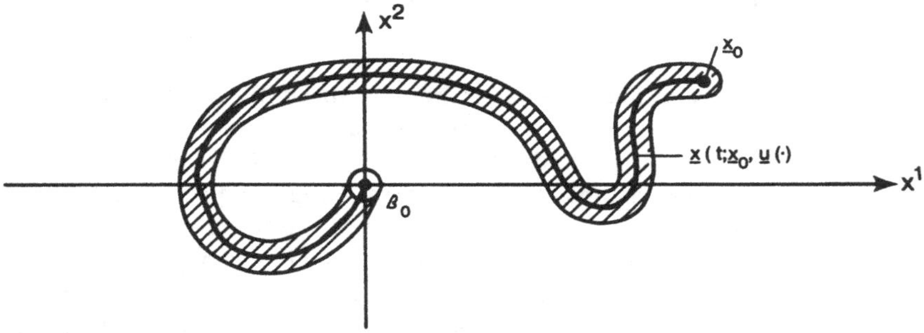

Figure 11

Because $\mathbf{f}(\mathbf{x}, \mathbf{u})$ and the entries of the Jacobian matrix $\mathbf{f}_\mathbf{x}(\mathbf{x}, \mathbf{u})$ are (uniformly) continuous on the compact set $\mathscr{N}$, there are constants $K > 0$, $\lambda \ge 0$, $\rho > 0$ such that on $\mathscr{N}$,

$$|\mathbf{f}(\mathbf{x}, \mathbf{y})| \le K, \qquad |\mathbf{f}(\mathbf{x}, \mathbf{u}) - \mathbf{f}(\mathbf{y}, \mathbf{u})| < \lambda |\mathbf{x} - \mathbf{y}|,$$

and

$$|\mathbf{u} - \mathbf{v}| < \rho \Rightarrow |\mathbf{f}(\mathbf{x}, \mathbf{u}) - f(\mathbf{x}, \mathbf{v})| < \frac{\delta e^{-\lambda t_1}}{4t_1}.$$

If S is a Lebesgue measurable set of real numbers, then $|S|$ will denote the Lebesgue measure of S. Now $\mathscr{U}_{PC}[0, t_1]$ is L^1-dense in $\mathscr{U}_m[0, t_1]$, which implies that it is dense in measure. This means that there is a measurable set $S \subset [0, t_1]$ and a $\mathbf{u}_\delta(\cdot) \in \mathscr{U}_{PC}$ such that $|S| < \delta e^{-\lambda t_1}/8K$, and

$$|\mathbf{u}(t) - \mathbf{u}_\delta(t)| < \rho \quad \text{if } t \in Q \equiv [0, t_1] \cap S^c.$$

We use $\chi_s(t)$ to denote the characteristic function of a set S. We can use (NLA) plus some judicious additions and subtractions to measure the difference between the $\mathscr{U}_{PC}$-response $\mathbf{x}_\delta[t] \equiv \mathbf{x}(t; \mathbf{x}_0, \mathbf{u}_\delta(\cdot))$ and the $\mathscr{U}_m$-response $\mathbf{x}[t] \equiv \mathbf{x}(t; \mathbf{x}_0, \mathbf{u}(\cdot))$:

$$\begin{aligned}|\mathbf{x}_\delta[t]-\mathbf{x}[t]| &\le \int_0^t |\mathbf{f}(\mathbf{x}_\delta[r], \mathbf{u}_\delta(r)) - \mathbf{f}(\mathbf{x}_\delta[r], \mathbf{u}(r))|[\chi_Q(r)+\chi_S(r)]\,dr \\ &\quad + \int_0^t |\mathbf{f}(\mathbf{x}_\delta[r], \mathbf{u}(r)) - \mathbf{f}(\mathbf{x}[r], \mathbf{u}(r))|\,dr \\ &\le \frac{\delta e^{-\lambda t_1}}{4} + 2K|S| + \lambda \int_0^t |\mathbf{x}_\delta[r]-\mathbf{x}[r]|\,dr.\end{aligned}$$

Using Gronwall's inequality (Hartman [1964], p. 24) we find that

$$|\mathbf{x}_\delta[t]-\mathbf{x}[t]| \le \left[\frac{\delta e^{-\lambda t_1}}{4} + 2K|S|\right] e^{\lambda t_1} < \left[\frac{\delta e^{-\lambda t_1}}{2}\right] e^{\lambda t_1} < \delta.$$

Therefore $\mathbf{x}_\delta(t_1) \in \mathscr{B}_0 \subset \mathscr{C}_{PC}$. This means that there is a control $\tilde{\mathbf{u}}(\cdot) \in \mathscr{U}_{PC}$, defined on $[0, \tilde{t}]$ that steers $\mathbf{x}_\delta(t_1)$ to $\mathbf{0}$, that is, the response $\tilde{\mathbf{x}}[t] \equiv \mathbf{x}(t; \mathbf{x}_\delta(t_1), \tilde{\mathbf{u}}(\cdot))$ satisfies $\tilde{\mathbf{x}}(\tilde{t}) = \mathbf{0}$. *Because our system is autonomous,* the following control from $\mathscr{U}_{PC}$ will then steer $\mathbf{x}_0$ to $\mathbf{0}$:

$$\mathbf{u}_0(t) = \begin{cases} \mathbf{u}_\delta(t), & 0 \le t \le t_1; \\ \tilde{\mathbf{u}}(t-t_1), & t_1 < t \le t_1 + \tilde{t}. \end{cases}$$

The associated response is $\mathbf{x}_\delta[t]$, $0 \le t \le t_1$, $\tilde{\mathbf{x}}(t-t_1)$ on $(t_1, t_1+\tilde{t}]$. □

The above proof is of course valid for any class $\tilde{\mathscr{U}}$ of controls that is

(1) dense in measure in $\mathscr{U}_m$, and satisfies
(2) for $\mathbf{u}_1(\cdot) \in \tilde{\mathscr{U}}[0, t_1]$, $\mathbf{u}_2 \in \tilde{\mathscr{U}}[0, t_2]$ the *juxtaposition*

$$\mathbf{u}_3(t) = \begin{cases} \mathbf{u}_1(t), & 0 \le t \le t_1; \\ \mathbf{u}_2(t-t_1), & t_1 < t \le t_2 + t_1 \end{cases}$$

is in $\tilde{\mathscr{U}}$. For example, the theorem holds if $\mathscr{U}_{PC}$ is replaced by $\tilde{\mathscr{U}} = \bigcup_{t_1>0} \tilde{\mathscr{U}}[0, t_1]$, where $\tilde{\mathscr{U}}[0, t_1]$ is the class of C^∞ functions which vanish, along with all derivatives, at 0 and t_1.

We now turn to controllability problems for $\mathscr{U}_{BB}$. This class is symmetric, but not convex, and since many of our proofs have depended heavily on convexity this class does not look promising. However, physical intuition tells us that $\mathscr{U}_{BB}$ should be effective, since we are using all available power. This intuition is correct for *linear* problems, as we have already mentioned in the discussion following the proof of Theorem 2. To conclude this chapter, we now re-state our special case of the Bang–Bang Principle. We give a proof based on ideas of Jim Yorke in the appendix to this chapter.

Proofs of the Bang–Bang Principle usually involve either a theorem of Liapunov on the range of a vector measure, or the Krein–Milman Theorem from functional analysis (cf., for example, H. Hermes and J. P. LaSalle [1969]). Our proof is long, but does not use such powerful tools.

Theorem 10 (The Bang–Bang Principle). *For* (LA), *we have* $\mathscr{C}_{BB}(t_1) = \mathscr{C}(t_1)$ *for all* $t_1 > 0$; *this set is compact, convex, and depends continuously on* t_1.

Exercises

Remark: Unless stated otherwise, we always assume that $|u^i(t)| < 1$, $i = 1, \ldots, m$.

1. (A proof of the lemma used in the proof of Theorem 7.) Consider the time-reversed autonomous system $\dot{\mathbf{x}} = -A\mathbf{x} - B\mathbf{u}$. Suppose that $\mathbf{0}$ can be steered to any point in a fixed ball $\mathscr{B}_0 = \mathscr{B}(\mathbf{0}; \delta)$. Then, given a set of scalar multiples of the canonical basis vectors,

$$S = \{\alpha \mathbf{e}_1, \alpha \mathbf{e}_2, \ldots, \alpha \mathbf{e}_n\}$$

with $\alpha > 0$ sufficiently small, there are controls $\mathbf{u}_1(\cdot), \mathbf{u}_2(\cdot), \ldots, \mathbf{u}_n(\cdot)$ which respectively steer $\mathbf{0}$ to $\alpha \mathbf{e}_1, \ldots, \alpha \mathbf{e}_n$ *at* $t_1 = 1$, i.e.,

$$\mathbf{x}(1; \mathbf{0}, \mathbf{u}_j(\cdot)) = \alpha \mathbf{e}_j, \qquad j = 1, 2, \ldots, n.$$

2. Consider the system $(n = 2, m = 1)$:

$$\mathbf{x} = \begin{bmatrix} p \\ q \end{bmatrix}, \qquad \dot{\mathbf{x}} = \begin{bmatrix} 0 & 1 \\ 0 & 0 \end{bmatrix}\mathbf{x} + \begin{bmatrix} 1 \\ 0 \end{bmatrix} u(t), \qquad \mathscr{T}(t) \equiv \mathbf{0},$$

a slight modification of the rocket car (compare the 2×1 matrices B). Use one of the theorems in this chapter to show that $\mathscr{C} \neq R^2$. Find $\mathscr{C}$ by direct integration.

3. (For $\mathbf{x}_0 \in \mathscr{C}$, the states lying on the successful response curve from $\mathbf{x}_0$ may not be in $\mathscr{C}$.)

   Consider the system $\dot{p} = -(1-t)q$, $\dot{q} = (1-t)p$ for $0 \le t < 1$, with $\dot{p} = 0$, $\dot{q} = [u(t) - 2](t-1)$ for $1 \le t < \infty$.

   Show that for $0 \le t < 1$, responses move counterclockwise on circles in the (p, q) plane, while for $t > 1$ responses move vertically downward. Describe $\mathscr{C}$, and sketch a typical successful response from $\mathscr{C}$. Show that *no* state on a successful response, except the initial state, is in $\mathscr{C}$.

4. (Although $\mathscr{C}(t_1) = \mathscr{C}_{BB}(t_1)$, the number of switches required may be infinite.)

   For the scalar problem $\dot{x} = [t^3 \sin(1/t)]u(t)$, $-1 \le u(t) \le 1$, $\mathscr{T}(t) \equiv 0$, show that $\mathscr{C}(1) = [-a, a]$, where

$$a = \int_0^1 t^3 \left|\sin\left(\frac{1}{t}\right)\right| dt.$$

Show that a is only attained by a control with an infinite number of switches.

5. Investigate the controllable set (with $\mathscr{T}(t) \equiv 0$) for the following systems.

$$\text{(a)}\ \dot{\mathbf{x}} = \begin{bmatrix} 0 & -1 & 1 \\ 2 & -3 & 1 \\ 1 & -1 & -1 \end{bmatrix} \mathbf{x} + \begin{bmatrix} -1 & 1 \\ 0 & 2 \\ 1 & 3 \end{bmatrix} \mathbf{u} \qquad (m = 2,\ n = 3)$$

$$\text{(b)}\ \dot{\mathbf{x}} = \begin{bmatrix} -p + 3[q]^2 + rq \\ -q - 3pq - rp \\ -r[p]^2 - r[q]^2 - [r]^3 \end{bmatrix} + \begin{bmatrix} -\alpha e^{\beta} + \beta \\ \alpha^2 + 2\beta \\ e^{\alpha} + e^{3\beta} \end{bmatrix},$$

with

$$\mathbf{x} = \begin{bmatrix} p \\ q \\ r \end{bmatrix}, \qquad \mathbf{u} = \begin{bmatrix} \alpha \\ \beta \end{bmatrix}.$$

6. Prove Theorem 5′.
Hint: It might be helpful to compare the controllable set for $\mathbf{u}(t) \in \Omega$ with the controllable set for $\mathbf{u}(t) \in k\Omega$, $k > 1$, $k\Omega \equiv \{k\mathbf{v} | \mathbf{v} \in \Omega\}$.

7. Consider the system ($m = 1$, $n = 3$):

$$\dot{\mathbf{x}} = \begin{bmatrix} 0 & 1 & 0 \\ 0 & -1 & 1 \\ 0 & 0 & -1 \end{bmatrix} \mathbf{x} + u(t)\mathbf{b}.$$

For which constant vectors $\mathbf{b}$ is $\mathscr{C} = R^n$?

8. Consider the nonlinear system ($m = n = 2$):

$$\dot{p} = qe^p - p - u^1,$$

$$\dot{q} = pe^q - q + u^2, \qquad \mathbf{x} = \begin{bmatrix} p \\ q \end{bmatrix}.$$

Show that $\mathscr{C} \neq R^2$.

9. Let (r, ϕ) be polar coordinates in the plane, and consider the control problem

$$\dot{r} = ur, \qquad \dot{\phi} = -u, \qquad u(\cdot) \in \mathscr{U}_1,$$

with $\phi(0) = 0$, $r(0) = 1$. Show that the reachable set in the plane is not convex. (It is a spiral.)

10. Let (r, ϕ) be polar coordinates in the plane and consider the control problem

$$\dot{r} = u^1 r, \qquad \dot{\phi} = -u^2, \qquad \mathbf{u}(\cdot) = \begin{bmatrix} u^1(\cdot) \\ u^2(\cdot) \end{bmatrix} \in \mathscr{U}_2$$

with initial state $r(0) = 1$, $\phi(0) = 0$. Show that $K(\pi; \mathbf{x}_0)$ is not convex. (It is an annulus.)

Appendix to Chapter II: Proof of the Bang–Bang Principle

The proof is long, so we have broken it down into a series of propositions. First we need a general compactness result.

Proposition 1 (Weak Sequential Compactness of $\mathscr{U}_m[0, t_1]$ in $L^2[0, t_1]$). *If $\{\mathbf{u}_n(\cdot)\} \subset \mathscr{U}_m[0, t_1]$ is given $(n = 1, 2, \ldots)$, then there exists a subsequence $\{\mathbf{u}_{n_k}(\cdot)\}$ and a $\mathbf{u}_*(\cdot) \in \mathscr{U}_m[0, t_1]$, such that for all $n \times m$ matrices $Y(t)$ whose entries are in $L^2[0, t_1]$, we have*

$$\lim_{k \to \infty} \int_0^{t_1} Y(t)\mathbf{u}_{n_k}(t)\, dt = \int_0^{t_1} Y(t)\mathbf{u}_*(t)\, dt.$$

Proof. The sequence of first components $\{u_n^1(\cdot)\}_{n=1}^{\infty}$ is a sequence in $L^2[0, t_1]$, since $u_n^1(\cdot)$ is measurable and satisfies $|u_n^1(t)| \leq 1$. Also if $\|\cdot\|_2$ is the $L^2[0, t_1]$ norm, then

$$\|u_n^1(\cdot)\|_2 = \left\{\int_0^{t_1} [u_n^1(t)]^2\, dt\right\}^{1/2} \leq t_1^{1/2}.$$

Now the ball in $L^2[0, t_1]$ about the origin of radius $t_1^{1/2}$ is weakly compact (cf. the Mathematical Appendix), therefore, there is a subsequence $\{u_{n_i}^1(\cdot)\}_{i=1}^{\infty}$ and a $u_*^1(\cdot) \in L^2[0, t_1]$ such that for all real-valued $f(\cdot) \in L^2[0, t_1]$,

$$\lim_{i \to \infty} \int_0^{t_1} f(t)u_{n_i}^1(t)\, dt = \int_0^{t_1} f(t)u_*^1(t)\, dt.$$

We throw away the original sequence $\{\mathbf{u}_n(\cdot)\} \subset \mathscr{U}_m$ and work with the new subsequence $\{\mathbf{u}_{n_i}(\cdot)\} \subset \mathscr{U}_m$, and for simplicity we relabel these functions $\{\mathbf{u}_n(\cdot)\}$. Beginning with this relabelled sequence, we now repeat the above procedure for the second component, then for the third, and so on. The end result is a subsequence $\{\mathbf{u}_k(\cdot)\} \subset \mathscr{U}_m$ and a $\mathbf{u}_*(\cdot) \in L^2[0, t_1]$ such that

$$\lim_{k \to \infty} \int_0^{t_1} f(t)\mathbf{u}_n(t)\, dt = \int_0^{t_1} f(t)\mathbf{u}_*(t)\, dt,$$

for all real-valued $f(\cdot) \in L^2[0, t_1]$, where we integrate componentwise.

We claim that $\mathbf{u}_*(\cdot) \in \mathscr{U}_m$, i.e., $|u_*^j(t)| \leq 1$ on $[0, t_1]$ for $j = 1, 2, \ldots, m$. Let $\mathscr{H} = \{t \mid \exists j,\ u_*^j(t) > 1\}$. If $\mathscr{H}$ has Lebesgue measure zero, $|\mathscr{H}| = 0$, then we can redefine $\mathbf{u}_*(t)$ to be zero on $\mathscr{H}$ without affecting anything, and thus $u_*(\cdot) \in \mathscr{U}_m$. So assume $|\mathscr{H}| > 0$ and for definiteness assume $\mathscr{H}_1 = \{t \mid u_*^1(t) > 1\}$ has positive measure. Then if $\chi(t)$ is the characteristic function of $\mathscr{H}_1$, we have (recall $\mathbf{u}_k(\cdot) \in \mathscr{U}_m \Rightarrow |u_k^1(t)| \leq 1$):

$$|\mathscr{H}_1| < \int_0^{t_1} \chi(t)u_*^1(t)\, dt = \lim_{k \to \infty} \int_0^{t_1} \chi(t)u_k^1(t)\, dt \leq \int_0^{t_1} \chi(t)\, dt = |\mathscr{H}_1|,$$

a contradiction. Thus $\mathbf{u}_*(\cdot) \in \mathscr{U}_m$. If $Y(t)$ is any $n \times m$ matrix with entries

$y_{ij}(\cdot)\in L^2[0,t_1]$, then $Y(t)\mathbf{u}_k(t)$ is a vector whose i^{th} component $[Y(t)\mathbf{u}_k(t)]^i$ satisfies

$$\lim_{k\to\infty}\int_0^{t_1}[Y(t)\mathbf{u}_k(t)]^i\,dt=\lim_{k\to\infty}\int_0^{t_1}\left(\sum_{j=1}^m y_{ij}u_k^j\right)dt=\sum_{j=1}^m\lim_{k\to\infty}\int_0^{t_1}y_{ij}u_k^j\,dt$$

$$=\sum_{j=1}^m\int_0^{t_1}y_{ij}u_*^j\,dt=\int_0^{t_1}[Y(t)\mathbf{u}_*(t)]^i\,dt. \qquad \square$$

To continue with the proof of Theorem 10, we set $m=1$, that is, $u(t)$ real-valued (the extension to $m>1$ is easy, working with each entry of $u(\cdot)$). Recall

$$\mathcal{U}_1[0,t_1]=\{u(\cdot)\mid -1\le u(t)\le 1,\ u(\cdot)\text{ measurable on }[0,t_1]\},$$
$$\mathcal{U}_{\mathbf{BB}}[0,t_1)=\{u(\cdot)\in\mathcal{U}_1[0,t_1]\mid |u(t)|=1\text{ for all }t\in[0,t_1]\}.$$

Our system is (LA) $\dot{\mathbf{x}}=A\mathbf{x}+Bu$, B an $n\times 1$ matrix, and

$$\mathscr{C}(t_1)=\{\mathbf{x}_0\in R^n\mid\exists u(\cdot)\in\mathcal{U}_1[0,t_1]\text{ steering }\mathbf{x}_0\text{ to }\mathbf{0}\},$$
$$\mathscr{C}_{\mathbf{BB}}(t_1)=\{\mathbf{x}_0\in R^n\mid\exists u(\cdot)\in\mathcal{U}_{\mathbf{BB}}[0,t_1]\text{ steering }\mathbf{x}_0\text{ to }\mathbf{0}\}.$$

We want to show that $\mathscr{C}(t_1)=\mathscr{C}_{\mathbf{BB}}(t_1)$, and that both sets are compact and convex. Now by Equation (3),

$$\mathbf{x}_0\in\mathscr{C}(t_1)\Leftrightarrow\mathbf{x}_0=-\int_0^{t_1}e^{-As}Bu(s)\,ds\text{ for some }u(\cdot)\in\mathcal{U}_1[0,t_1],$$

$$\mathbf{x}_0\in\mathscr{C}_{\mathbf{BB}}(t_1)\Leftrightarrow\mathbf{x}_0=-\int_0^{t_1}e^{-As}Bu(s)\,ds\text{ for some }u(\cdot)\in\mathcal{U}_{\mathbf{BB}}[0,t_1].$$

Therefore, if T is the mapping from $\mathcal{U}_1[0,t_1]$ into R^n defined by

$$(10)\quad T(u(\cdot))=\int_0^{t_1}\mathbf{y}(s)u(s)\,ds,\qquad \mathbf{y}(s)\equiv e^{-As}B\ (\text{recall that }B\text{ is }n\times 1),$$

then we need to prove that

$$(11)\qquad \mathrm{T}(\mathcal{U}_1[0,t_1])\equiv Q=Q_{\mathbf{BB}}\equiv T(\mathcal{U}_{\mathbf{BB}}[0,t_1]).$$

The remaining propositions are devoted to the proof of (11). Notice that $Q_{\mathbf{BB}}\subset Q$ always, since $\mathcal{U}_{\mathbf{BB}}\subset\mathcal{U}$. For simplicity we set $t_1=1$.

Proposition 2. *The mapping T defined in* (10) *maps convex sets into convex sets, and weakly compact sets in $L^2[0,1]$ into compact sets in R^n.* (*In particular, by Proposition* 1, *Q is convex and compact.*)

Proof. T is linear, so the convexity assertion is immediate.

It is a standard result in functional analysis that a mapping which takes weakly convergent sequences into strongly convergent sequences will map weakly (sequentially) compact sets into (sequentially) compact sets (cf.

Mathematical Appendix). Since $\mathscr{U}_m[0, 1]$ is weakly compact by Proposition 1, we need only show that the mapping $T[u(\cdot)]$ has the above property. Let $\{u_n(\cdot)\} \subset \mathscr{U}_1[0, 1]$ converge weakly to $u(\cdot) \in \mathscr{U}_1[0, 1]$, that is, for all $f \in L^2[0, 1]$,

$$\lim_{n\to\infty} \int_0^1 f(t)u_n(t)\,dt = \int_0^1 f(t)u(t)\,dt.$$

Then by looking at components we see that $T(u_n) \to T(u)$. Thus $Q = T(\mathscr{U}_1[0, 1])$ is compact. □

We now come to the heart of the proof of the Bang–Bang Principle. For G, H disjoint measurable subsets of $[0, 1]$ ($G \cup H$ need not be $[0, 1]$) we define the special control classes (Figure 12):

$$\mathscr{U}(G, H) \equiv \{u(\cdot) \in \mathscr{U}_1[0, 1] | u(t) = -1 \text{ on } G,\ u(t) = +1 \text{ on } H\},$$

$$\mathscr{U}_{\mathrm{BB}}(G, H) \equiv \mathscr{U}_{\mathrm{BB}}[0, 1] \cap \mathscr{U}(G, H).$$

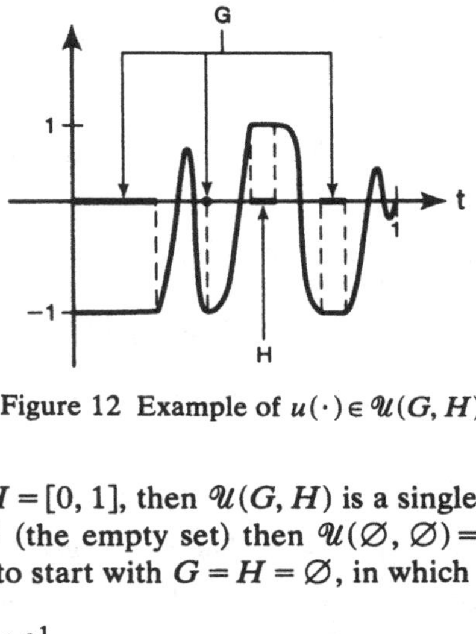

Figure 12 Example of $u(\cdot) \in \mathscr{U}(G, H)$

Notice that if $G \cup H = [0, 1]$, then $\mathscr{U}(G, H)$ is a single bang–bang control, while if $G = H = \varnothing$ (the empty set) then $\mathscr{U}(\varnothing, \varnothing) = \mathscr{U}_1[0, 1]$. The basic idea of the proof is to start with $G = H = \varnothing$, in which case

$$T[\mathscr{U}(\varnothing, \varnothing)] = \left\{\int_0^1 \mathbf{y}(s)u(s)\,ds \middle| \mathbf{y}(s) = e^{-As}B,\ u(\cdot) \in \mathscr{U}_1[0, 1]\right\} \equiv Q.$$

Now if $\mathbf{p} \in Q = T[\mathscr{U}(\varnothing, \varnothing)]$, we will systematically choose larger and larger sets G, H so that $\mathbf{p}$ remains in the convex set $T[\mathscr{U}(G, H)]$ (cf. Figure 13), but the dimension of $T[\mathscr{U}(G, H)]$ drops. Eventually we get $\mathbf{p} \in T[\mathscr{U}(G, H)]$ with $\dim T[\mathscr{U}(G, H)] = 0$. Since $T[\mathscr{U}(G, H)]$ is convex, it must be a single point, so $\{\mathbf{p}\} = T[\mathscr{U}(G, H)] \supset T[\mathscr{U}_{\mathrm{BB}}(G, H)] \neq \varnothing$. Thus $\mathbf{p} = T[\mathscr{U}_{\mathrm{BB}}(G, H)]$, so $\mathbf{p} \in Q_{\mathrm{BB}}$.

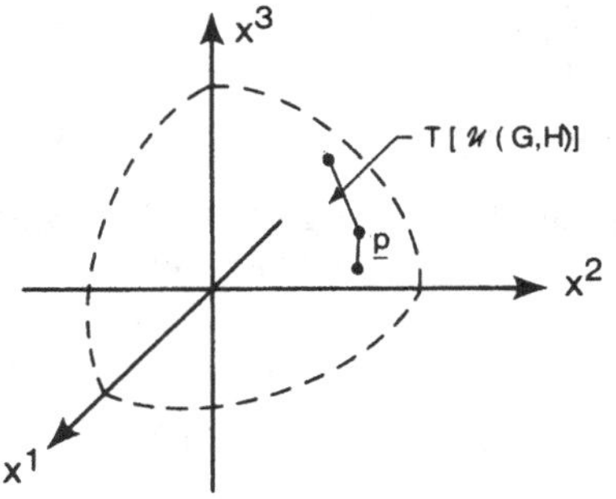

Figure 13 dim $Q = 3$, dim $T[\mathscr{U}(G, H)] = 1$

Proposition 3. *$T[\mathscr{U}(G, H)]$ is always compact and convex in R^n.*

Proof

(i) $\mathscr{U}(G, H)$ is convex and $T[\cdot]$ is linear on $\mathscr{U}(G, H)$, so $T[\mathscr{U}(G, H)]$ is convex.

(ii) $\mathscr{U}_1[0, 1]$ is weakly compact (Proposition 1) and $T[\cdot]$ maps weakly convergent sequences into strongly convergent sequences. If we can show that $\mathscr{U}(G, H) \subset \mathscr{U}_1[0, 1]$ is always weakly closed, then it will follow from standard results (cf. Mathematical Appendix) that $\mathscr{U}(G, H)$ is weakly compact and $T[\mathscr{U}(G, H)]$ is compact. To this end, let $\{u_n(\cdot)\} \subset \mathscr{U}(G, H)$. Then we may assume that this sequence converges weakly (in $L^2[0, 1]$) to $u(\cdot) \in \mathscr{U}_1[0, 1]$. We want to show that $u(\cdot) \in \mathscr{U}(G, H)$, i.e., that $u(t) \equiv -1$ on G, $u(t) \equiv +1$ on H. Suppose that there is a subset $E \subset G$, $|E| > 0$ such that $u(t) \neq -1$ on all E (if $|E| = 0$, we can redefine $u(t)$ to be -1 on E without affecting the response). Since $u(\cdot) \in \mathscr{U}_1[0, 1]$, we have $u(t) > -1$ on E, so

$$-|E| < \int_E u(t)\,dt = \int_0^1 \chi_E(t) u(t)\,dt.$$

Now $u_n(\cdot) \in \mathscr{U}(G, H)$ for each n, so $u_n(t) \equiv -1$ on E, therefore as $n \to \infty$,

$$-|E| = \int_0^1 \chi_E(t) u_n(t)\,dt \to \int_0^1 \chi_E(t) u(t) > -|E|.$$

This contradiction shows that $\mathscr{U}(G, H)$ is weakly closed and so $T[\mathscr{U}(G, H)]$ is compact. □

To finish the proof, we need to establish two final propositions. We state them here, and immediately use them to finish the proof of the Bang–Bang Principle. Their proofs are given after. Recall that for a convex set $K \subset R^n$, dim K is the dimension of the smallest hyperplane containing K.

Proposition 4. *If $\mathbf{p} \in \operatorname{Int} T[\mathscr{U}(G, H)]$, then there exist disjoint measurable subsets of $[0, 1]$, $G^* \supset G$, $H^* \supset H$ such that $\mathbf{p} \in \partial T[\mathscr{U}(G^*, H^*)]$.*

Proposition 5. *If* $\mathbf{p}\in\partial T[\mathscr{U}(G^*,H^*)]$, *then there are disjoint measurable subsets of* $[0, 1]$, $G^{**}\supset G^*$, $H^{**}\supset H^*$ *such that* $\mathbf{p}\in T[\mathscr{U}(H^{**}, G^{**})]$ *and*

$$\dim T[\mathscr{U}(G^{**}, H^{**})] < \dim T[\mathscr{U}(G^*, H^*)].$$

To complete the proof of the Bang–Bang Principle using these results, let $\mathbf{p}\in Q=T(\mathscr{U}_1[0,1])=T[\mathscr{U}(\varnothing,\varnothing)]$ be arbitrary. Applying Propositions 4 and 5 alternately, starting with $H_1=G_1=\varnothing$, we obtain increasing sequences of sets $\varnothing=G_1\subset G_2\subset\cdots$, $\varnothing=H_1\subset H_2\subset\cdots$ with

$$n\geq\dim T[\mathscr{U}(G_k,H_k)]>\dim T[\mathscr{U}(G_{k+1},H_{k+1})],$$

so $\dim T[\mathscr{U}(G_k,H_k)]$ drops by at least one at each step. After at most n such steps, we will have two disjoint measurable sets G, H such that $\dim T[\mathscr{U}(G,H)]=0$ and $\mathbf{p}\in T[\mathscr{U}(G,H)]$. But $T[\mathscr{U}(G,H)]$ is convex, therefore $T[\mathscr{U}(G,H)]=\{\mathbf{p}\}$. Also $\mathscr{U}_{\mathrm{BB}}(G,H)$ is nonempty and contained in $\mathscr{U}(G,H)$, so $\varnothing\neq T[\mathscr{U}_{\mathrm{BB}}(G,H)]\subset T[\mathscr{U}(G,H)]=\{\mathbf{p}\}$. Thus $\{\mathbf{p}\}=T[\mathscr{U}_{\mathrm{BB}}(G,H)]\subset T(\mathscr{U}_{\mathrm{BB}}[0,1])\equiv Q_{\mathrm{BB}}$. This shows that $Q\subset Q_{\mathrm{BB}}$, so $Q_{\mathrm{BB}}=Q$. □

Proof of Proposition 4. Fix G and H in $[0,1]$, $G\cap H=\varnothing$. Let $\mathbf{p}\in\operatorname{Int}T[\mathscr{U}(G,H)]$. For $s\in[0,1]$, define

$$G_s=G\cup\{[0,s]-H\},\qquad H_s=H,$$

(so that we can "increase G smoothly without hitting H"). Since $T[\mathscr{U}(G_s,H_s)]$ is a compact subset of R^n, we can use the Hausdorff metric on the family of sets $\{T[\mathscr{U}(G_s,H_s)]0\leq s\leq1\}$ in R^n. For fixed sets G, H, we claim that the set-valued map

$$W(\cdot)\colon s\to T[\mathscr{U}(G_s,H_s)]$$

is continuous on $[0,1]$. To see this, let $s,t\in[0,1]$, and without loss of generality let $s\leq t$, so $G_s\subset G_t$ (as always, $H=H_s=H_t$). Then $W(s)\subseteq W(t)$ and the Hausdorff distance can be computed by

$$h(W(s),W(t))=\sup_{\mathbf{p}\in W(s)}d(\mathbf{p},W(t)),$$

where $d(\mathbf{p},W(t))=\inf_{\mathbf{q}\in W(t)}|\mathbf{p}-\mathbf{q}|$. We will show that for any $\mathbf{p}\in W(s)$, there is a $\mathbf{q}\in W(t)$ such that $|\mathbf{p}-\mathbf{q}|<K|t-s|$, for some fixed constant $K>0$, and therefore $h(W(s),W(t))<K(t-s)$ and the continuity of $W(\cdot)$ follows. Given $\mathbf{p}\in W(s)$, i.e.,

$$\mathbf{p}=\int_0^1\mathbf{y}(\tau)u_s(\tau)\,d\tau\quad\text{for some } u_s(\cdot)\in\mathscr{U}(G_s,H_s),$$

where $\mathbf{y}(\tau)=e^{-A\tau}B$, we define $u_t(\cdot)\in\mathscr{U}(G_t,H_t)$:

$$u_t(\tau)=\begin{cases}-1 & \text{on } G_t,\\ 1 & \text{on } H_t=H,\\ u_s(\tau) & \text{on } (G_t\cup H)^c.\end{cases}$$

Then $u_s(\tau) \neq u_t(\tau)$ only for $\tau \in (G_t - G_s) \subset [s, t]$, and so

$$\left| \int_0^1 \mathbf{y}(\tau) u_s(\tau)\, d\tau - \int_0^1 \mathbf{y}(\tau) u_t(\tau)\, d\tau \right| \leq 2\left(\max_{[0,1]} |\mathbf{y}(\tau)| \right) |t - s|.$$

Therefore $W(s) = T[\mathscr{U}(G_s, H_s)]$ is a continuous map from $[0, 1]$ into the subsets of R^n.

Given $\mathbf{p} \in \operatorname{Int} T[\mathscr{U}(G, H)]$, we define $\sigma = \sup\{s \in [0, 1] \| \mathbf{p} \in T[\mathscr{U}(G_s, H_s)]\}$. This set of numbers is nonempty ($s = 0$ belongs), so σ exists. From continuity and the definition of σ, $\mathbf{p} \notin \operatorname{Int} T[\mathscr{U}(G_\sigma, H_\sigma)]^c$. On the other hand, because $T[\mathscr{U}(G_s, H_s)]$ is continuous in s, $\mathbf{p} \notin \operatorname{Int} T[\mathscr{U}(G_\sigma, H_\sigma)]$. Thus $\mathbf{p} \in \partial T[\mathscr{U}(G_\sigma, H_\sigma)]$. □

Proof of Proposition 5. This proof is similar to the proof of Pontryagin's Maximum Principle for general linear systems. Recall that $\mathbf{x} \in T[\mathscr{U}(G, H)]$ means

$$\mathbf{x} = \int_0^1 \mathbf{y}(s) u(s)\, ds \quad \text{for some } u(\cdot) \in \mathscr{U}(G, H),$$

i.e., $u(\cdot) \equiv -1$ on G, $u(\cdot) \equiv +1$ on H. Let $\mathbf{p} \in \partial T[\mathscr{U}(G, H)]$; by Proposition 3, $\mathbf{p} \in T[\mathscr{U}(G, H)]$, so $\mathbf{p} = \int_0^1 \mathbf{y}(\tau) u_p(\tau)\, d\tau$ for some $u_p(\cdot) \in \mathscr{U}(G, H)$. Since $T[\mathscr{U}(G, H)]$ is convex, there exists a unit vector $\mathbf{q} \in R^k$, $|\mathbf{q}| = 1$ such that (cf. Mathematical Appendix)

$$\int_0^1 \mathbf{q}^T \mathbf{y}(\tau) u_p(\tau)\, d\tau \equiv \mathbf{q}^T \mathbf{p} = \sup\{\mathbf{q}^T \mathbf{x} | \mathbf{x} \in T[\mathscr{U}(G, H)]\}$$

(12) $$\equiv \sup\left\{ \int_0^1 \mathbf{q}^T \mathbf{y}(\tau) v(\tau)\, d\tau | v(\cdot) \in \mathscr{U}(G, H) \right\},$$

$$\mathbf{y}(\tau) = e^{-A\tau} B.$$

Now to maximize $\int_0^1 \mathbf{q}^T \mathbf{y}(\tau) v(\tau)\, d\tau$, we want $\operatorname{sgn} v(\tau) = \operatorname{sgn} \mathbf{q}^T \mathbf{y}(\tau)$ on $[0, 1]$. With this in mind, we define

$$G_* = G \cup (\{t | \mathbf{q}^T \mathbf{y}(t) < 0\} - H), \qquad H_* = H \cup (\{t | \mathbf{q}^T \mathbf{y}(t) > 0\} - G),$$

$$D = \{\mathbf{x} \in T[\mathscr{U}(G, H)] \| \mathbf{q}^T \mathbf{x} = \mathbf{q}^T \mathbf{p}\}.$$

The set D is the intersection in R^k of the k-dimensional set $T[\mathscr{U}(G, H)]$ with the hyperplane $\{\mathbf{x} | \mathbf{q}^T(\mathbf{x} - \mathbf{p}) = 0\}$. This hyperplane in turn is a translate of the subspace $\{\mathbf{x} | \mathbf{q}^T \mathbf{x} = 0\}$ which has dimension $k - 1$. Thus $\dim D \leq k - 1$, i.e., $\dim D < \dim T[\mathscr{U}(G, H)]$.

We will show that $D = T[\mathscr{U}(G_*, H_*)]$. The fact that $D \supset T[\mathscr{U}(G_*, H_*)]$ is easy to verify. If $\mathbf{x}_0 \in D \subset T[\mathscr{U}(G, H)]$, then

$$\mathbf{q}^T \mathbf{x}_0 = \int_0^1 \mathbf{q}^T \mathbf{y}(\tau) u_0(\tau)\, d\tau \quad \text{for some } u_0(\cdot) \in \mathscr{U}(G, H).$$

We will show that $u_0(\cdot)\in\mathcal{U}(G_*, H_*)$. From the definition of D and (12) we have

$$(13)\quad \int_0^1 \mathbf{q}^T\mathbf{y}(\tau)u_0(\tau)\,d\tau \equiv \mathbf{q}^T\mathbf{x}_0 = \sup\left\{\int_0^1 \mathbf{q}^T\mathbf{y}(\tau)v(\tau)\,d\tau \,\middle|\, v(\cdot)\in\mathcal{U}(G,H)\right\}.$$

Clearly, then, $\operatorname{sgn} u_0(\tau) = \operatorname{sgn} \mathbf{q}^T\mathbf{y}(\tau)$, so $u_0(\cdot)\in\mathcal{U}(G_*, H_*)$. Thus $\mathbf{x}_0\in D \Rightarrow \mathbf{x}_0\in T[\mathcal{U}(G_*, H_*)]$ so $D = T[\mathcal{U}(G_*, H_*)]$. □

The extension of the above proofs to the case $m>1$ is relatively straightforward. In this case we have a continuous matrix $Y(\tau)\equiv e^{-A\tau}B$ defined on $[0,1]$, and $T[\cdot]$: $\mathcal{U}_m[0,1]\to R^n$ is defined by

$$T[\mathbf{u}(\cdot)] = \int_0^1 Y(s)\mathbf{u}(s)\,ds.$$

For two families of subsets of $[0,1]$, $\mathbf{G}=\{G^1, G^2, \ldots, G^m\}$, $\mathbf{H}=\{H^1, H^2, \ldots, H^m\}$, with $G^i\cap H^i=\varnothing$, $i=1,2,\ldots,m$, we define

$$\mathcal{U}(\mathbf{G},\mathbf{H})\equiv\{\mathbf{u}(\cdot)\in\mathcal{U}_m[0,1] \mid u^i(t)=-1 \text{ on } G^i,\ u^i(t)=+1 \text{ on } H^i,\ i=1,2,\ldots,m\},$$

and we "extend **G** without hitting **H**" by defining

$$G_s^i = G^i\cup\{[0,s]-H^i\},\qquad H_s^i = H^i,\qquad i=1,2,\ldots,m.$$

Finally, we set $\sigma=\sup\{s \mid \mathbf{p}\in T[\mathcal{U}(\mathbf{G}_s,\mathbf{H}_s)]\}$, and if the ith column of $Y(t)$ is denoted $\mathbf{y}_i(t)$, we define

$$G_*^i = G^i\cup(\{t \mid \mathbf{q}^T\mathbf{y}_i(t)<0\}-H^i),\ H_*^i = H^i\cup\{t \mid \mathbf{q}^T\mathbf{y}_i(t)>0\}.$$ □

Chapter III

Linear Autonomous Time-Optimal Control Problems

1. Introduction: Summary of Results

In this chapter we will give a complete treatment of the linear optimal control problem:

(LA) $\quad \dot{\mathbf{x}}(t) = A\mathbf{x}(t) + B\mathbf{u}(t), \quad \mathbf{x}(t) \in R^n, \quad \mathbf{u}(t) \in \Omega \subset R^m,$

with A, B constant $n \times n$ and $n \times m$ matrices, $\mathbf{u}(\cdot) \in \mathscr{U}_m$, target $\mathscr{T}(t) \equiv \mathbf{0}$, and cost function

$$C[\mathbf{u}(\cdot)] = \int_0^{t_1} 1 \, dt = t_1,$$

where t_1 is the time of arrival at the target $\mathbf{0}$. A successful control and its response are *time-optimal* if there is no successful control which gives a smaller value of t_1. There may be many time-optimal controls for a given initial state.

We assume *throughout this chapter* that

no column of B consists entirely of zeros.

This assumption involves no loss of generality. If, for example, the first column of B is all zeros, then the first component $u^1(\cdot)$ of $\mathbf{u}(\cdot)$ is irrelevant in (LA). Therefore, we may replace the m-vector $\mathbf{u}(\cdot)$ by the $(m-1)$-vector $[u_2(\cdot), u_3(\cdot), \ldots, u_m(\cdot)]^T$. This eliminates the first column of B in (LA) without changing the problem.

Many results from this chapter are true for the general linear system

$$\dot{\mathbf{x}}(t) = A(t)\mathbf{x}(t) + B(t)\mathbf{u}(t) + \mathbf{c}(t),$$

and we have summarized these extensions in the final section. We are devoting an entire chapter to the specific problem (LA) because the geometry of the situation is much more clear than in the case of a general optimal control problem. The intuition gained from an understanding of this geometry should help in Chapters IV and V.

First we give an outline of the results to be proved in this chapter. To do this, recall that Ω is the unit cube in R^m:

$$\Omega=\{\mathbf{v}|\mathbf{v}\in R^m, |v^i|\leq 1, i=1,2,\ldots,m\};$$

that $K(t;\mathbf{x}_0)$ is the reachable set from $\mathbf{x}_0$ at time t:

$$K(t;\mathbf{x}_0)=\{\mathbf{x}(t;\mathbf{x}_0,\mathbf{u}(\cdot))|\mathbf{u}(\cdot)\in\mathcal{U}_m\};$$

that $\mathrm{RC}=\bigcup_{t\geq 0}\{t\}\times K(t;\mathbf{x}_0)$ is the reachable cone, and that a bang–bang control $\mathbf{u}(\cdot)$ always satisfies $|u^i(t)|=1$, $i=1,2,\ldots,m$. Notice that $K(t;\mathbf{x}_0)$ is a subset of R^n and RC is a subset of R^{n+1} (cf. Figures 4 and 5 of Chapter I).

$K(t,\mathbf{x}_0)$ and RC may be imbedded in hyperplanes of minimal dimension in R^n and R^{n+1} respectively. Then Int K, ∂K, etc., are all taken relative to the appropriate hyperplane. For the autonomous system (LA), given an initial state $\mathbf{x}_0\in R^n$, *suppose that there exists a successful control* from $\mathcal{U}_m$ steering $\mathbf{x}_0$ to $\mathbf{0}$ (not necessarily time-optimal). Then the following propositions hold:

(a) There exists at least one bang–bang time-optimal control in $\mathcal{U}_m$ (not necessarily piecewise constant).

(b) The response to any time-optimal control lies on the boundary of $K(t;\mathbf{x}_0)$ for all t, i.e., if $\mathbf{x}[t]\equiv\mathbf{x}(t;\mathbf{x}_0,\mathbf{u}(\cdot))$ is time-optimal, then $\mathbf{x}[t]\in\partial K(t;\mathbf{x}_0)$ (equivalently, $(t,\mathbf{x}[t])\in\partial\mathrm{RC}$) for $0\leq t\leq t_1$.

(c) A time-optimal control $\mathbf{u}(t)$ satisfies the *maximum principle*: there is a constant vector $\mathbf{h}\neq\mathbf{0}$ such that

$$\mathbf{h}^T e^{-At}B\mathbf{u}(t)=\sup_{\mathbf{v}\in\Omega}\mathbf{h}^T e^{-At}B\mathbf{v},\qquad 0\leq t\leq t_1;$$

equivalently,

$$u^i(t)=\operatorname{sgn}(\mathbf{h}^T e^{-At}B)^i,\qquad i=1,2,\ldots,m,$$

whenever $(\mathbf{h}^T e^{-At}B)^i\neq 0$.

(d) Under a certain *normality* condition, the time-optimal control is unique, bang–bang, and piecewise constant.

(e) If the normality condition holds, then the converse of (c) is valid: any successful control which satisfies the maximum principle is in fact time-optimal (hence, by (d), unique, bang–bang, and piecewise constant).

Remark. We will also show that the number of switches depends in certain cases on the eigenvalues of A.

To begin, we need to make some preliminary observations. Recall that the Bang–Bang Principle (Theorem 10 from Chapter II) states that for *any* linear autonomous system, $\mathscr{C}_{BB}(t) = \mathscr{C}(t)$ for all $t > 0$, where $\mathscr{C}(t)$ ($\mathscr{C}_{BB}(t)$) is the set of initial states for which there exists a successful control from $\mathscr{U}_m$ ($\mathscr{U}_{BB}$) steering the system to the target state $\mathbf{0}$. Note also that we choose $\mathscr{T}(t) \equiv \mathbf{0}$ for convenience – any fixed point in R^n could be the target. We claim that the Bang–Bang Principle is equivalent to the following statement about the reachable set:

For (LA), $K(t; \mathbf{x}_0) = K_{BB}(t; \mathbf{x}_0)$ for all $t \geq 0$, and all $\mathbf{x}_0$.

That is, the states which can be reached from a given $\mathbf{x}_0$ at a given time are the same whether we use general controls from $\mathscr{U}_m[0, t]$, or bang–bang controls.

This follows from the idea of *time-reversal* discussed in Chapter II. Recall that $\mathbf{x}(t)$ solves (LA) with $\mathbf{x}(0) = \mathbf{x}_0$ if and only if $\mathbf{z}(t) = \mathbf{x}(t_1 - t)$ solves the time-reversed system

$$(*) \qquad \dot{\mathbf{z}} = -A\mathbf{z} - B\mathbf{u}, \qquad \mathbf{z}(t_1) = \mathbf{x}_0.$$

Therefore the two systems have the same curves as trajectories, traversed in opposite directions. The time-reversed system is again linear and autonomous, so the Bang-Bang Principle asserts that $\mathscr{C}_{BB}(t_1) = \mathscr{C}(t_1)$ for (*) – that is, the states steerable to $\mathbf{x}_0$ under (*) are the same for bang–bang as for general controls. But $\mathbf{x}_1$ can be steered to $\mathbf{x}_0$ under (*) if and only if $\mathbf{x}_0$ can be steered to $\mathbf{x}_1$ under (LA) (Figure 1). Thus $\mathbf{x}_1$ is in the controllable set for (*) if and only if it is in the reachable set for (LA). Therefore the Bang–Bang Principle implies that $K_{BB}(t_1; \mathbf{x}_0) = K(t_1; \mathbf{x}_0)$ for (LA).

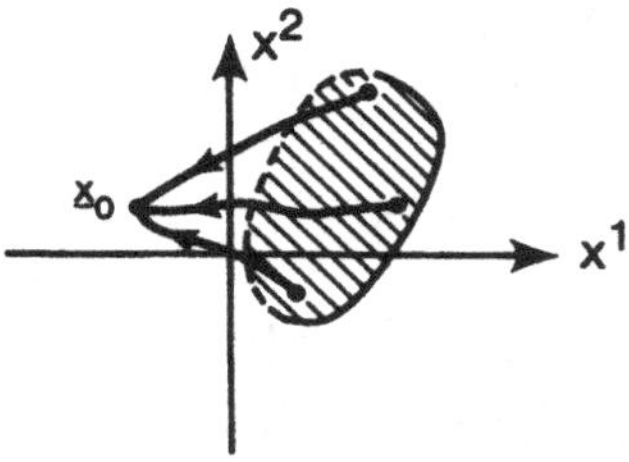

$\mathscr{C}(t_1)$ for (LA) (with target $\mathbf{x}_0$)

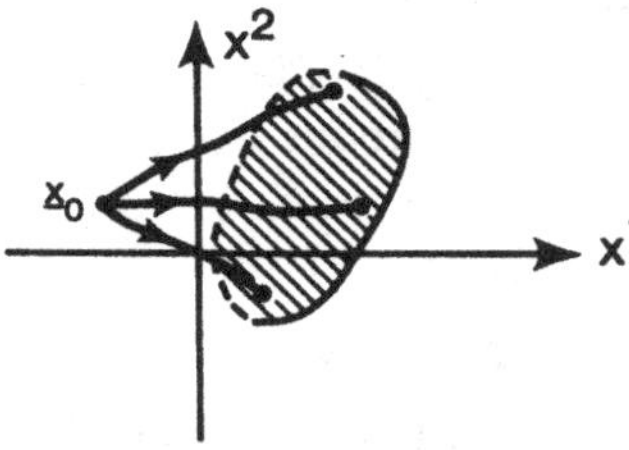

$K(t_1, \mathbf{x}_0)$ for (*)

Figure 1

We also need the following technical lemma, which is just the remainder of the statement of the Bang–Bang Principle (Theorem 10) from Chapter II.

Lemma. *For (LA), $K(t; \mathbf{x}_0)$ is always convex and compact; if $\mathbf{x}_0 = \mathbf{0}$ it is also symmetric. Furthermore, the set-valued mapping*

$$t \to K(t; \mathbf{x}_0), \qquad 0 \leq t < \infty,$$

is continuous (using the Hausdorff metric in the range).

Proof. The response formula for (LA) is (compare Equation (1) from Chapter II)

$$\mathbf{x}(t; \mathbf{x}_0, \mathbf{u}(\cdot)) = e^{At}\mathbf{x}_0 + \int_0^t e^{A(t-s)}B\mathbf{u}(s)\,ds \equiv e^{At}\mathbf{x}_0 + Q[\mathbf{u}].$$

Therefore $\mathbf{y} \in K(t; \mathbf{x}_0)$ if and only if $\mathbf{y} = e^{At}\mathbf{x}_0 + Q[\mathbf{u}]$ for some $\mathbf{u}(\cdot) \in \mathscr{U}_m[0, t]$. For t fixed, the convexity of $K(t; \mathbf{x}_0)$ (and the symmetry of $K(t, \mathbf{0})$) follow immediately from the linearity of Q and the fact that $\mathscr{U}_m[0, t]$ is symmetric and convex. Proposition 1 of the Appendix to Chapter II states that $\mathscr{U}_m[0, t]$ is weakly sequentially compact in $L^2[0, t]$, and Proposition 2 of that Appendix shows that for each t, the set

$$\mathscr{D}(t) = \left\{ \int_0^t e^{-As}B\mathbf{u}(s)\,ds \,\middle|\, \mathbf{u}(\cdot) \in \mathscr{U}_m[0, t] \right\}$$

is compact in R^n. The set $K(t, \mathbf{x}_0)$ is obtained from this set by multiplying each element by e^{At} and adding the fixed vector by $e^{At}\mathbf{x}_0$ (both continuous operations), hence $K(t, \mathbf{x}_0)$ is compact in R^n.

Finally, we will show that the map $t \to \mathscr{D}(t)$ is continuous. For fixed t_1 and any $\varepsilon > 0$, we must show that there is a $\delta(\varepsilon) > 0$ such that $|t_* - t_1| < \delta$ implies that $\mathscr{D}(t_1)$, $\mathscr{D}(t_*)$ are each contained in an ε-sack about the other. Let us agree that $\delta < 1$, and set $T = t_1 + 1$, $M = \max_{[0,T]} |e^{-As}B|$. We will show that $\mathscr{D}(t_*)$ is contained in an ε-sack about $\mathscr{D}(t_1)$ for $|t_* - t_1| < \varepsilon/M$. Let $\mathbf{y}_* \in \mathscr{D}(t_*)$, i.e., $\mathbf{y}_* = \int_0^{t_*} e^{-As}B\mathbf{u}_*(s)\,ds$ for some $\mathbf{u}_*(\cdot) \in \mathscr{U}_m[0, t_*]$. Extend $\mathbf{u}_*(s)$ to $[0, T]$ by setting it equal to zero on $(t_*, T]$, then define $\mathbf{y}_0 = \int_0^{t_1} e^{-As}B\mathbf{u}_*(s)\,ds$. Clearly $\mathbf{y}_0 \in \mathscr{D}(t_1)$ and $|\mathbf{y}_* - \mathbf{y}_0| \leq |\int_{t_*}^{t_1} M\,ds| < \varepsilon$. Thus $\mathbf{y}_*$ is in the ε-sack about $\mathscr{D}(t_1)$. □

2. The Existence of a Time-Optimal Control; Extremal Controls; the Bang–Bang Principle

Theorem 1. *If there exists a successful control steering $\mathbf{x}_0$ to $\mathbf{0}$ under (LA) then there exists a time-optimal control.*

Proof. Our assumption that there is a successful control means that $\mathbf{0} \in K(t_*; \mathbf{x}_0)$ for some t_*. Let

$$t_1 = \inf\{t \geq 0 \mid \mathbf{0} \in K(t; \mathbf{x}_0)\}.$$

This set is non-empty and bounded below, so the infimum exists. We claim $\mathbf{0} \in K(t_1; \mathbf{x}_0)$, which will imply that there is a control in $\mathcal{U}_m$ steering $\mathbf{x}_0$ to $\mathbf{0}$ in the minimal time t_1.

Suppose $\mathbf{0} \in K(t_1; \mathbf{x}_0)$. Since $K(t_1; \mathbf{x}_0)$ is closed there is a ball $\mathcal{B}(\mathbf{0}; \rho)$ about $\mathbf{0} \in R^n$ such that $\mathcal{B}(\mathbf{0}; \rho) \cap K(t_1; \mathbf{x}_0) = \varnothing$. Since $K(t; \mathbf{x}_0)$ is continuous in t, we can preserve this empty intersection for t near t_1 if we shrink $\mathcal{B}(\mathbf{0}; \rho)$, i.e., there is a $\delta > 0$ such that

$$\mathcal{B}(\mathbf{0}; \rho/2) \cap K(t; \mathbf{x}_0) = \varnothing \quad \text{for } t_1 \le t \le t_1 + \delta.$$

This says that $\mathbf{0} \in R^n$ is *not* reachable for $t_1 \le t \le t_1 + \delta$, contradicting the definition of t_1. □

Corollary. *If there exists a successful control from $\mathcal{U}_m$ steering $\mathbf{x}_0$ to $\mathbf{0}$, then there exists a bang–bang time-optimal control.*

Proof. By the Bang–Bang Principle, $K_{\mathrm{BB}}(t; \mathbf{x}_0) = K(t; \mathbf{x}_0)$ for each $t \ge 0$. □

Definition (Figure 2). A control $\mathbf{u}(\cdot)$ defined on $[0, t_*]$ is *extremal* if the associated response lies on the boundary of RC, i.e.,

(1) $$\mathbf{x}(t; \mathbf{x}_0, \mathbf{u}(\cdot)) \in \partial K(t; \mathbf{x}_0), \qquad 0 \le t \le t_*.$$

An extremal control may or may not be successful, and if successful, it may or may not be optimal.

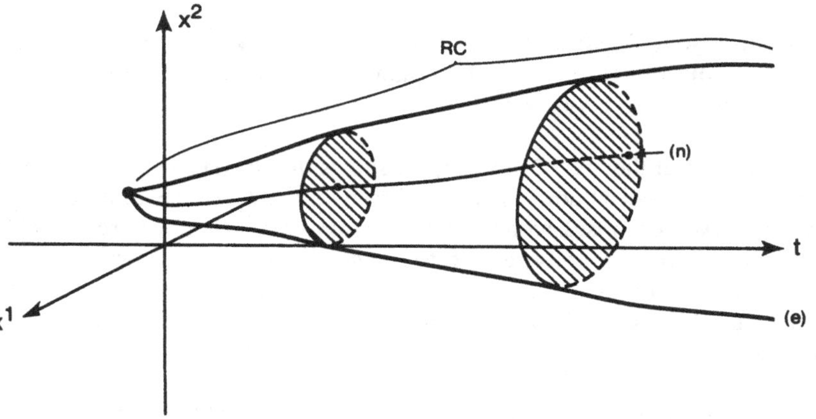

Figure 2 The Curve (e) Is Extremal, (n) Is Not

Theorem 2. *If $\mathbf{w}(\cdot)$ is a time-optimal control for (LA), then $\mathbf{w}(\cdot)$ is extremal.*

Proof. The proof is in two parts. First, we will show that if $\mathbf{w}(\cdot)$ is optimal, then at the instant t_1 of arrival at $\mathbf{0}$, the response will lie on $\partial K(t_1; \mathbf{x}_0)$. Second, we will show that if a response lies on $\partial K(t_*; \mathbf{x}_0)$ at any fixed instant t_*, then (1) holds for $0 \le t \le t_*$. (In other words, a response can never move from the interior of RC to the boundary, ∂RC.)

For the first part of the proof, assume that $\mathbf{w}(\cdot)$ is an optimal control, steering $\mathbf{x}_0$ to $\mathbf{0}$ in time t_1, i.e.,

$$\mathbf{x}[t_1] \equiv \mathbf{x}(t_1; \mathbf{x}_0, \mathbf{w}(\cdot)) = \mathbf{0},$$

and suppose that $\mathbf{x}[t_1] = \mathbf{0}$ is not on $\partial K(t_1; \mathbf{x}_0)$. Then there is a ball $\mathscr{B}(\mathbf{0}; \rho) \subset K(t_1; \mathbf{x}_0)$. Because $K(t_1; \mathbf{x}_0)$ is a continuous function of t, we can vary t and still preserve the above inclusion if we shrink the ball, i.e., there is a $\delta > 0$ such that $\mathscr{B}(\mathbf{0}; \rho/2) \subset K(t; \mathbf{x}_0)$ for $t_1 - \delta \le t \le t_1$. Then $\mathbf{0}$ would be attainable at time $t_1 - \delta$, contradicting the optimality of t_1. Therefore, $\mathbf{0} = \mathbf{x}[t_1] \in \partial K(t_1; \mathbf{x}_0)$.

Turning to the second part of the proof, suppose that $\mathbf{x}_* \equiv \mathbf{x}[t_*] \in \operatorname{Int} K(t_*; \mathbf{x}_0)$ for some $0 < t_* < t_1$, where $\mathbf{x}[t]$ is the response to $\mathbf{w}(\cdot)$. We must show that $\mathbf{x}[t] \in \operatorname{Int} K(t; \mathbf{x}_0)$ for all $t > t_*$. Since $\mathbf{x}_* \in \operatorname{Int} K(t_*; \mathbf{x}_0)$, there is a ball $\tilde{B} \equiv B(\mathbf{x}_*; \delta) \subset K(t_*; \mathbf{x}_0)$. Thus each state $\tilde{\mathbf{x}}_0 \in \tilde{B}$ can be reached from $\mathbf{x}_0$ at time t_*, using some admissible control $\tilde{\mathbf{u}}(\cdot)$. We consider the new problem

$$\dot{\mathbf{y}} = A\mathbf{y} + B\mathbf{w}, \qquad \mathbf{y}(t_*) = \tilde{\mathbf{x}}_0, \qquad t_* < t$$

with *fixed* control $\mathbf{w}(\cdot)$. Note that for $\tilde{\mathbf{x}}_0 = \mathbf{x}_*$, we have $\mathbf{y}[t] \equiv \mathbf{x}[t]$. The solution of this problem can be written

$$\mathbf{y}[t] = e^{A(t-t_*)}\tilde{\mathbf{x}}_0 + \int_{t_*}^{t} e^{A(t-s)} B\mathbf{w}(s)\, ds = R(t)\tilde{\mathbf{x}}_0 + \mathbf{c}(t).$$

This one-to-one linear map from $\tilde{B}$ into $K(t; \mathbf{x}_0)$ takes open sets into open sets, thus the image of the open sphere $\tilde{B}$ lies in $\operatorname{Int} K(t; \mathbf{x}_0)$. Thus the image of $\mathbf{x}_*$, which is just $\mathbf{x}[t]$, lies in $\operatorname{Int} K(t; \mathbf{x}_0)$. □

Exercise 1 at the end of this chapter shows that the converse of Theorem 2 is false – we need a "normality" condition for the converse. Since extremal controls give responses which lie on a boundary, it is not surprising that they are in fact intimately related to bang–bang controls, representing the maximum use of available power. This is the content of the following theorem, which states that "normally," extremal controls are bang–bang controls.

Theorem 3. *Let* $\mathbf{u}(\cdot) \in \mathscr{U}_m[0, t_*]$. *Then* $\mathbf{u}(\cdot)$ *is extremal for* (LA) *if and only if there is a nonzero constant vector* $\mathbf{h}$ *such that for* $0 \le t \le t_*$,

(2a) $$u^i(t) = \operatorname{sgn}\{\mathbf{h}^T e^{-At} B\}^i, \qquad i = 1, \ldots, m.$$

Furthermore, for each t, $0 \le t \le t_*$,

(2b) $$\mathbf{h}^T e^{-At} B\mathbf{u}(t) = \sup_{\mathbf{v} \in \Omega} \mathbf{h}^T e^{-At} B\mathbf{v}.$$

Proof. Suppose that $\mathbf{u}_e(t)$ is extremal on $[0, t_*]$, so $\mathbf{x}_e[t] \equiv \mathbf{x}(t; \mathbf{x}_0, \mathbf{u}_e(\cdot)) \in \partial K(t; \mathbf{x}_0)$, $0 \le t \le t_*$. Since $K(t_*; \mathbf{x}_0)$ is convex, there is a supporting hyperplane P through $\mathbf{x}_e[t_*]$ (Figure 3) such that $K(t_*; \mathbf{x}_0)$ lies on one side of P. Let $\mathbf{n}$ be such that (cf. the Mathematical Appendix)

(3) $$\langle \mathbf{n}, \mathbf{x}_e[t_*]\rangle = \sup \{\langle \mathbf{n}, \mathbf{x}\rangle | \mathbf{x} \in K(t_*; \mathbf{x}_0)\}.$$

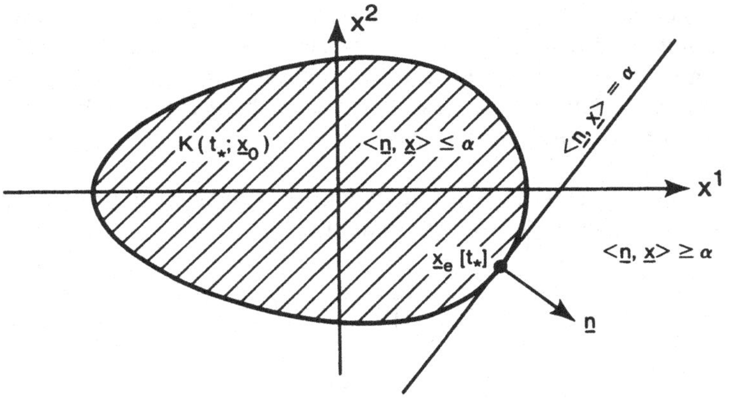

Figure 3

By the response formula, $\mathbf{x} \in K(t_*; \mathbf{x}_0)$ if and only if

$$\mathbf{x} = \mathbf{x}_0 e^{At_*} + \int_0^{t_*} e^{A(t_*-s)} B\mathbf{u}(s)\, ds, \qquad \mathbf{u}(\cdot) \in \mathcal{U}_m[0, t_*],$$

so (3) is equivalent to the following (the terms involving $\mathbf{x}_0$ cancel):

(4)

$$\mathbf{n}^T \int_0^{t_*} e^{A(t_*-s)} B\mathbf{u}_e(s)\, ds = \sup\left\{\mathbf{n}^T \int_0^{t_*} e^{A(t_*-s)} B\mathbf{u}(s)\, ds \,\middle|\, u(\cdot) \in \mathcal{U}_m[0, t_*]\right\}.$$

Define $\mathbf{h}^T = \mathbf{n}^T e^{At_*}$, and $Y(s) = e^{-As}B$. Then

(*) $$\int_0^{t_*} \mathbf{h}^T Y(s)\mathbf{u}_e(s)\, ds = \sup\left\{\int_0^{t_*} \mathbf{h}^T Y(s)\mathbf{u}(s)\, ds \,\middle|\, \mathbf{u}(\cdot) \in \mathcal{U}_m\right\},$$

and $\mathbf{h} \ne \mathbf{0}$ since e^{At_*} is non-singular. Since

$$\int_0^{t_*} \mathbf{h}^T Y(s)\mathbf{u}(s)\, ds = \sum_{i=1}^m \int_0^{t_*} \{\mathbf{h}^T Y(s)\}^i u^i(s)\, ds,$$

it follows that the sup in (*) is attained for $u_e^i(s) = \operatorname{sgn}\{\mathbf{h}^T Y(s)\}^i$, $i = 1, \ldots, m$, $0 \le t \le t_*$, and then $\mathbf{h}^T e^{-As} B\mathbf{u}_e(s) = \sup_{\mathbf{v}\in\Omega} \mathbf{h}^T e^{-As} B\mathbf{v}$. Conversely, if we choose $\mathbf{u}_e(\cdot)$ according to (2a), and define $\mathbf{n}$ by $\mathbf{n}^T = \mathbf{h}^T e^{-At_*}$, then (4) holds, which is equivalent to (3). But (3) implies that $\mathbf{x}_e[t_*]$ must lie on the boundary of $K(t_*; \mathbf{x}_0)$ (cf. the Mathematical Appendix).

Corollary (The Maximum Principle for (LA)). *If* $\mathbf{u}(\cdot)$ *is optimal then there exists an* $\mathbf{h} \neq \mathbf{0}$ *such that* (2a) *holds, and*

$$\mathbf{h}^T e^{-At} B\mathbf{u}(t) = \sup_{\mathbf{v}\in\Omega} \mathbf{h}^T e^{-At} B\mathbf{v}.$$

Remark. Each component of $\mathbf{h}^T e^{-At}B$ is an analytic function of t; therefore on a compact interval $[0, t_1]$ it is either identically zero, or vanishes at most a finite number of times. Therefore, the right side of Equation (2) either makes sense for all t, except for a finite (and unimportant) set of values of t, or it is undefined for *all* t. In this second case, the theorem tells us nothing about the corresponding component of $\mathbf{u}(t)$.

EXAMPLE 1 (The Rocket Car). The dynamics is given by $\dot{p}(t) = q(t)$, $\dot{q}(t) = u(t)$ (q = velocity, p = position). In system form,

$$A = \begin{bmatrix} 0 & 1 \\ 0 & 0 \end{bmatrix}, \qquad B = \begin{bmatrix} 0 \\ 1 \end{bmatrix},$$

$$e^{-At} = I - At = \begin{bmatrix} 1 & -t \\ 0 & 1 \end{bmatrix}$$

$$e^{-At}B = \begin{bmatrix} -t \\ 1 \end{bmatrix}.$$

By our previous work (Example 3 of Chapter II), $\mathscr{C} = R^2$. If $\mathbf{h}^T = [\alpha, \beta]$, then $\mathbf{h}^T e^{-At}B = \beta - \alpha t$.

The corollary above says that any optimal control $u(t)$ must satisfy $u(t) = \operatorname{sgn}(\beta - \alpha t)$ where α, β are not both zero. Thus the optimal control from any initial state is bang–bang with at most one switch.

EXAMPLE 2 (The Linearized Pendulum). The angular displacement $\theta(t)$ of a free pendulum from the vertical satisfies $\ddot{\theta}(t) + \sin\theta(t) = 0$. If we linearize about $\theta(t) \equiv 0$, and also add a forcing term, we have $\ddot{\theta} + \theta = u$. Our target is $\theta(t_1) = \dot{\theta}(t_1) = 0$, i.e., we want to bring the pendulum to rest. In system form,

$$\mathbf{x}(t) = \begin{bmatrix} \theta(t) \\ \dot{\theta}(t) \end{bmatrix}, \qquad \dot{\mathbf{x}} = \begin{bmatrix} 0 & 1 \\ -1 & 0 \end{bmatrix}\mathbf{x} + u(t)\begin{bmatrix} 0 \\ 1 \end{bmatrix}.$$

Then

$$e^{-At} = \begin{bmatrix} \cos t & -\sin t \\ \sin t & \cos t \end{bmatrix}, \qquad e^{-At}B = \begin{bmatrix} -\sin t \\ \cos t \end{bmatrix}.$$

If $\mathbf{h}^T = [\alpha, \beta]$, then

$$\mathbf{h}^T e^{-At}B = \beta\cos t - \alpha\sin t = \rho\sin(t + \delta),$$

where $\rho=[\alpha^2+\beta^2]^{1/2}>0$, $\delta=\arccos(-\alpha/\rho)$. So an optimal control $u(t)$ satisfies $u(t)=\operatorname{sgn}[\sin(t+\delta)]$, which means that it is bang–bang and periodic with period π. We will see later that the optimal control does *not* always strictly oppose the motion of the pendulum.

3. Normality and the Uniqueness of the Optimal Control

EXAMPLE 3 (An Indefinite Case). Consider the two-dimensional system $\dot{\mathbf{x}}(t)=\mathbf{u}(t)$ $(m=n=2)$. Then

$$A=\begin{bmatrix}0&0\\0&0\end{bmatrix},\qquad B=\begin{bmatrix}1&0\\0&1\end{bmatrix},\qquad e^{-At}=I,\qquad \mathbf{u}(\cdot)=\begin{bmatrix}u^1\\u^2\end{bmatrix},$$

and $\mathbf{h}^T e^{-At}B=\mathbf{h}^T\equiv(\alpha,\beta)$. Thus

$$(*)\qquad u^1(t)=\operatorname{sgn}\alpha,\qquad u^2(t)=\operatorname{sgn}\beta.$$

If either of α,β is zero, then (*) does not tell us anything about the corresponding component of $\mathbf{u}(t)$. In the preceding two examples, we didn't have to know $\mathbf{h}^T$ in order to reach certain qualitative conclusions; in this example we need more information.

To study this example more closely, consider the initial point $(-1,0)$ (Figure 4). This state can be steered to the target $(0,0)$ on the darkly sketched curve by the control

$$u_*(t)\equiv\begin{bmatrix}1\\0\end{bmatrix},$$

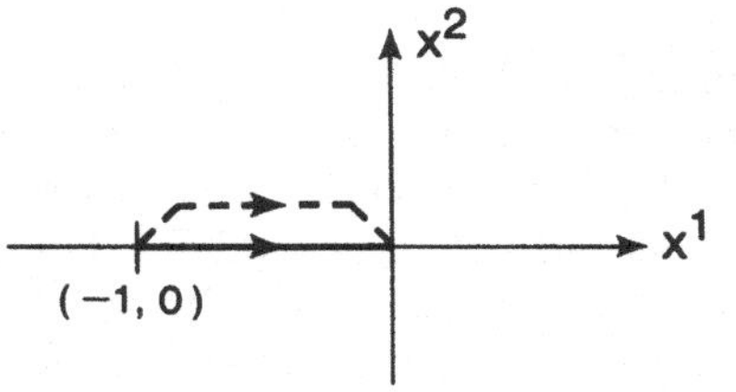

Figure 4

in the optimal time $t_1=1$. The time is optimal because we cover the shortest distance using maximum velocity. This is not a bang–bang control, and we must have in this case $\alpha>0$, $\beta=0$. Actually there are infinitely many optimal controls steering from $(-1,0)$ to $(0,0)$, e.g., $u^1(t)\equiv1$, and

$$u^2(t)=\begin{cases}1, & 0\le t\le a,\\ 0, & a<t<1-a,\\ -1, & 1-a\le t\le 1.\end{cases}$$

(The response is the dashed curve in Figure 4.) The idea is to "bump" $x^2(t)$, and then bring it back to zero. This control will be bang–bang in exactly one case ($a=\frac{1}{2}$). Interestingly, there are infinitely many optimal bang–bang controls – just keep bumping $x^2(t)$ up and down, instead of leaving it constant for $a<t<1-a$.

Analytically, in this example the difficulty stems from the fact that $\mathbf{h}^T e^{-At}B$ has a component *identically* zero; geometrically, the problem is that $K(t, \mathbf{x}_0)$ has "flat spots" on its boundary (cf. Exercise 2).

Definition. We call (LA) *normal* if for each constant vector $\mathbf{h}\neq\mathbf{0}$, no component of $\mathbf{h}^T e^{-At}B$ can vanish on a set of positive measure (equivalently, no component can vanish identically).

Of course this definition is of little practical use. We will eventually show that (LA) is normal if and only if the vectors $\{\mathbf{b}_j, A\mathbf{b}_j, \ldots, A^{n-1}\mathbf{b}_j\}$ are linearly independent for $j=1,2,\ldots,m$, where $\mathbf{b}_j$ is the jth column of B. This is an analytical characterization. Geometrically, (LA) is normal if and only if $K(t;\mathbf{x}_0)$ is *strictly* convex for all t, for one (hence all) $\mathbf{x}_0$. (The response formula (1) from Chapter II:

$$\mathbf{y}\in K(t;\mathbf{x}_0)\Leftrightarrow \mathbf{y}=e^{At}\mathbf{x}_0+\int_0^t e^{A(t-s)}B\mathbf{u}(s)\,ds$$

for some $\mathbf{u}(\cdot)\in U_m$, shows that the convexity of $K(t,\mathbf{x}_0)$ is independent of $\mathbf{x}_0$.)

Theorem 4. *If (LA) is normal, and if there exists a successful control (steering* $\mathbf{x}_0$ *to* $\mathbf{0}$*), then there exists a unique time-optimal control, which is bang–bang and piecewise constant.*

Proof. Existence follows from Theorem 1. By Theorem 3, normality implies that any optimal control must be bang–bang. If $\mathbf{u}(t)$ and $\mathbf{v}(t)$ were two distinct time-optimal bang–bang controls, then (because (LA) is linear) $\mathbf{w}(t)=\frac{1}{2}[\mathbf{u}(t)+\mathbf{v}(t)]$ would also be time-optimal, but not bang–bang. This contradiction to Theorem 3 implies that the optimal control is unique. Finally, the unique time-optimal bang–bang control is piecewise constant, since each of its components can only change sign when the corresponding component of $\mathbf{h}^T e^{-At}B$ is zero – as remarked earlier, each of these components is analytic in t, so it can only vanish a finite number of times for $0\leq t\leq t_1$. □

Example 4. (Uniqueness Does Not Imply Normality). Consider the two-dimensional system

$$\dot{\mathbf{x}}(t)=\begin{bmatrix}1 & 1\\ 1 & 1\end{bmatrix}\mathbf{u}(t),\qquad A=\begin{bmatrix}0 & 0\\ 0 & 0\end{bmatrix},\qquad B=\begin{bmatrix}1 & 1\\ 1 & 1\end{bmatrix}.$$

We compute

$$e^{-At}B=\begin{bmatrix}1 & 1\\ 1 & 1\end{bmatrix},\qquad \mathbf{h}^T e^{-At}B=\begin{bmatrix}\alpha+\beta\\ \alpha+\beta\end{bmatrix}^T,\qquad \mathbf{h}=\begin{bmatrix}\alpha\\ \beta\end{bmatrix}.$$

Thus $u^1(t)\equiv u^2(t)\equiv \operatorname{sgn}(\alpha+\beta)$, by Theorem 3, and so if $\alpha+\beta\neq 0$, there will be at least one optimal control which is bang–bang. Notice that the problem is not normal (e.g., $\mathbf{h}^T=[1,-1]$).

We can use Theorem 3 of Chapter II – we compute the controllability matrix:

$$M=[B, AB]=\begin{bmatrix}1 & 1 & 0 & 0\\ 1 & 1 & 0 & 0\end{bmatrix}.$$

Rank $M=1$, so the controllable set $\mathscr{C}$ is not all of R^2. Actually, a direct (and straightforward) investigation of the differential equation involved shows that $\mathscr{C}=\{(x^1,x^2)|x^1=x^2\}$. Consider the initial state $\mathbf{x}_0=(-1,-1)$. For this $\mathbf{x}_0$, $K(t;\mathbf{x}_0)$ is a line segment of length $4t$, centered at $(-1,-1)$, on the line $\mathscr{C}$ (Figure 5). Note that *any* vector $\mathbf{h}^T=(\alpha,\beta)$ normal to $K(t;\mathbf{x}_0)$ has $\alpha+\beta=0$. Clearly, the optimal time is $t_1=\frac{1}{2}$, with optimal control $u^1(t)\equiv u^2(t)\equiv 1$. This optimal control is clearly unique, and it is bang–bang, in spite of the fact that the problem is not normal.

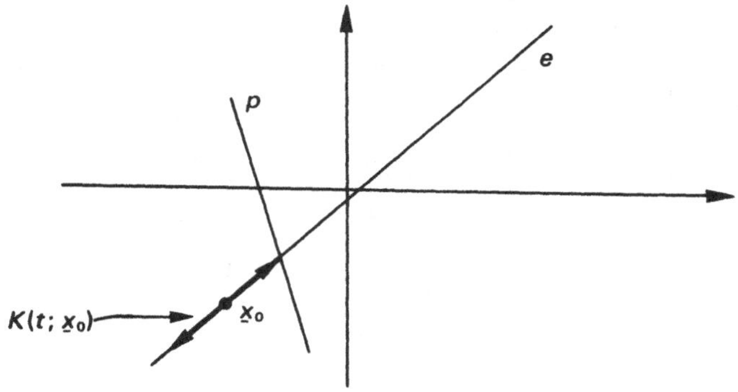

Figure 5

Note that in this example, $K(t;\mathbf{x}_0)$ is contained in a one-dimensional hyperplane, so ∂K and Int K are taken relative to this hyperplane:

$$\operatorname{Int} K=\{(p,p)|-1-2t<p<-1+2t\},$$
$$\partial K=(-1-2t,-1-2t)\cup(-1+2t,-1+2t).$$

In the preceding example, we had uniqueness of the optimal control in spite of non-normality. Notice that the target is an extreme point of $K(t;\mathbf{x}_0)$ at the optimal time $t=\frac{1}{2}$. Exercise 2 (a continuation of Example 3) shows that when $\mathbf{0}$ is not an extreme point, the optimal control may not be unique. Theorem 5 below shows that this is no accident. In order to prove this theorem, we first must investigate the uniqueness of successful responses. It often happens that there are many controls steering from $\mathbf{x}_0$ to the target. However the associated responses may all coincide.

Definition. Let $\mathbf{x}_0$ be given. We say that the *response from* $\mathbf{x}_0$ to a point $\mathbf{y}$ in $K(t_*; \mathbf{x}_0)$ *is unique* if every control which steers from $\mathbf{x}_0$ to $\mathbf{y}$ in time t_* generates the same response function, i.e., if $\mathbf{u}(\cdot)$, $\mathbf{v}(\cdot)$ are successful controls for $0 \le t \le t_*$, then $\mathbf{x}(t; \mathbf{x}_0, \mathbf{u}(\cdot)) \equiv \mathbf{x}(t; \mathbf{x}_0, \mathbf{v}(\cdot))$ on $[0, t_*]$.

There are two types of uniqueness we need to keep in mind. The type defined above says that there may be many time-optimal controls, but they all generate the same response function. Then there is the idea of a unique time-optimal control, which of course would generate a single response. The lemma below shows that these concepts are identical: we have a unique response function if and only if we have a unique time-optimal control steering $\mathbf{x}_0$ to $\mathbf{y}$.

EXAMPLE 5. With $m = n = 2$, we consider the system from Example 4. Then $\mathbf{u}(t) \equiv [1, 0]^T$ and $\mathbf{v}(t) = [0, 1]^T$ both steer the state $(-1, -1)$ to $\mathbf{0}$ at the time $t_1 = 1$, with the same response: $\mathbf{x}[t] = (-1+t)[\frac{1}{1}]$. A consequence of the following lemma is that in this case there must be a third control which steers $(-1, -1)$ to $\mathbf{0}$ in unit time with a *different* response function. For example (Figure 6):

$$\mathbf{w}(t) = \begin{cases} \begin{bmatrix} 1 \\ 1 \end{bmatrix}, & 0 \le t \le \frac{3}{4}; \\ \begin{bmatrix} -1 \\ -1 \end{bmatrix}, & \frac{3}{4} < t \le 1. \end{cases}$$

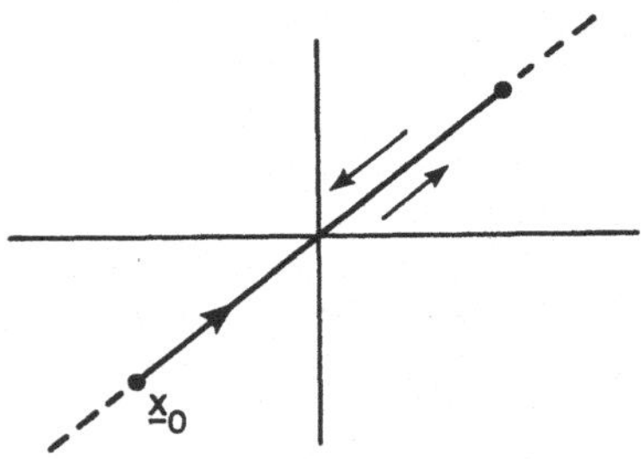

Figure 6

The proof of this lemma is the first place we need our (nonrestrictive) assumption that $\mathbf{b}_j \neq \mathbf{0}$ for all columns $\mathbf{b}_j$ of B.

Lemma. *Assume that B does not contain a column of zeros. Let $\mathbf{y} \in K(t_*; \mathbf{x}_0)$. Then the control steering $\mathbf{x}_0$ to $\mathbf{y}$ at time t_* is unique if and only if the response function from $\mathbf{x}_0$ to $\mathbf{y}$ in time t_* is unique.*

Proof. If the successful control is unique, then of course there is only one response.

On the other hand, suppose there are two controls $\mathbf{u}_1(\cdot)$, $\mathbf{u}_2(\cdot)$ which steer $\mathbf{x}_0$ to $\mathbf{y}$ in time t_*. Without loss of generality assume $\mathbf{u}_1(\mathbf{x})$ is bang–bang. We will show that there are at least two distinct response functions. One (or more) of the three controls

$$\mathbf{u}_1(\cdot), \quad \mathbf{u}_2(\cdot), \qquad \tfrac{1}{2}[\mathbf{u}_1(\cdot)+\mathbf{u}_2(\cdot)]$$

is *not* bang–bang – call it $\mathbf{u}_0(t)$. Then (since (LA) is linear), $\mathbf{u}_0(\cdot)$ steers $\mathbf{x}_0$ to $\mathbf{y}$ in time t_* and some fixed component $u_0^j(t)$ satisfies $|u_0^j(t)|<1$ on a set $S \subset [0, t_*]$ of positive measure. The response formula (Equation (1) from Chapter II) for (LA) states that

$$\mathbf{y} = e^{At_*}\left\{\mathbf{x}_0 + \int_0^{t_*} e^{-As}B\mathbf{u}_0(s)\,ds\right\}. \tag{5}$$

Now $\mathbf{y}$ is attained at time t_*, i.e., (5) holds, with $\mathbf{u}_0(\cdot)$ replaced by $\mathbf{u}_1(\cdot)$, so

$$\int_0^{t_*} e^{-As}B\mathbf{u}_0(s)\,ds = \int_0^{t_*} e^{-As}B\mathbf{u}_1(s)\,ds.$$

But this is not quite good enough for our purposes, so we have to be a bit more subtle. We claim that there is a *scalar* bang–bang function $\phi(\cdot)$ such that

$$\int_0^{t_*} e^{-As}\mathbf{b}_j u_0^j(s)\,ds = \int_0^{t_*} e^{-As}\mathbf{b}_j\phi(s)\,ds, \tag{6}$$

where $\mathbf{b}_j$ is the jth column of B. This stems directly from the response formula and the Bang–Bang Principle in one (control) dimension, applied to the following control problem:

$$\dot{\mathbf{x}}(t) = A\mathbf{x}(t) + \mathbf{b}_j u(t) + \mathbf{c}(t) \qquad (m=1), \qquad \mathbf{c}(t) = \tilde{B}\tilde{\mathbf{u}}_0(t)$$

where the $n \times (m-1)$ matrix $\tilde{B}$ is obtained from B by removing the jth column $\mathbf{b}_j$, and $\tilde{\mathbf{u}}_0(t)$ is the fixed $(m-1)$-vector $\mathbf{u}_0(t)$ with $u_0^j(t)$ removed. The forcing term $\mathbf{c}(t)$ doesn't affect the validity of the Bang–Bang Principle and the response formula will change in an obvious way.

All of the above is a precise but complicated way of describing the control problem obtained from (LA) by *fixing* each component of $\mathbf{u}(\cdot)$, except the jth, to be the corresponding (given) component of $\mathbf{u}_0(\cdot)$, thus leaving $u^j(\cdot)$ as our control, and reducing the dimension of the control space to one. Then the response formula and the Bang–Bang Principle (for a one-dimension control source) give us (6).

We now define a new m-dimensional control $\mathbf{w}(\cdot)$ for our original problem by fixing $w^i(t) = u_0^i(t)$ for $i \neq j$, and using $w^j(t) = \phi(t)$ as control. Then $\mathbf{w}(\cdot)$ steers $\mathbf{x}_0$ to $\mathbf{y}$ in time t_*.

We will show that the response functions $\mathbf{x}(t; \mathbf{x}_0, \mathbf{u}_0(\cdot))$ and $\mathbf{x}(t; \mathbf{x}_0, \mathbf{w}(\cdot))$, $0 \leq t \leq t_*$, are not identical. We suppose that

$$\mathbf{x}(t; \mathbf{x}_0, \mathbf{w}(\cdot)) = \mathbf{x}(t; \mathbf{x}_0, \mathbf{u}_0(\cdot)), \qquad 0 \leq t \leq t_*,$$

and look for a contradiction. This equality implies that

$$\int_0^t e^{-As} B\mathbf{w}(s)\, ds = \int_0^t e^{-As} B\mathbf{u}_0(s)\, ds, \qquad 0 \leq t \leq t_*$$

and so $B[\mathbf{w}(t) - \mathbf{u}_0(t)] = \mathbf{0}$ a.e. on $[0, t_*]$ (Exercise 5). But $\mathbf{w}(t) - \mathbf{u}_0(t)$ has only the jth component nonzero, so $\mathbf{b}^j[\phi(t) - u_0^j(t)] = \mathbf{0}$ a.e. But $|\phi(t)| = 1$ for all t, while $|u_0^j(t)| < 1$ on a set of positive measure, hence $\mathbf{b}^j = \mathbf{0}$, a contradiction to our basic assumption that B does not have any column of zeros. □

Theorem 5. *Suppose that* $\mathbf{y} \in K(t_*; \mathbf{x}_0)$. *Then the control steering from* $\mathbf{x}_0$ *to* $\mathbf{y}$ *at time* t_* *is unique if and only if* $\mathbf{y}$ *is an extreme point of* $K(t_*; \mathbf{x}_0)$.

Proof. Suppose that there are two controls $\mathbf{u}_1(\cdot)$, $\mathbf{u}_2(\cdot)$ steering $\mathbf{x}_0$ to $\mathbf{y}$ in time t_* with distinct responses, i.e., for at least one t_c, $0 < t_c < t_*$,

$$\mathbf{z}_1 \equiv \mathbf{x}(t_c; \mathbf{x}_0, \mathbf{u}_1(\cdot)) \neq \mathbf{x}(t_c; \mathbf{x}_0, \mathbf{u}_2(\cdot)) \equiv \mathbf{z}_2,$$

$$\mathbf{z}_i = e^{At_c}\mathbf{x}_0 + \int_0^{t_c} e^{A(t_c - s)} B\mathbf{u}_i(s)\, ds, \qquad i = 1, 2.$$

We "cut, interchange, and splice" $\mathbf{u}_1(\cdot)$ and $\mathbf{u}_2(\cdot)$ to form new controls

$$\mathbf{v}(t) = \begin{cases} \mathbf{u}_1(t), & 0 \leq t \leq t_c, \\ \mathbf{u}_2(t), & t_c < t \leq t_*. \end{cases} \qquad \mathbf{w}(t) = \begin{cases} \mathbf{u}_2(t), & 0 \leq t \leq t_c, \\ \mathbf{u}_1(t), & t_c < t \leq t_*. \end{cases}$$

We will show that the states $\mathbf{p} \equiv \mathbf{x}(t_*; \mathbf{x}_0, \mathbf{v}(\cdot))$ and $\mathbf{q} \equiv \mathbf{x}(t_*; \mathbf{x}_0, \mathbf{w}(\cdot))$ are distinct and $\mathbf{y} = \frac{1}{2}(\mathbf{p} + \mathbf{q})$, hence $\mathbf{y}$ is not an extreme point of $K(t_*; \mathbf{x}_0)$.

Now

$$\mathbf{y} = e^{At_*}\mathbf{x}_0 + \int_0^{t_*} e^{A(t_* - s)} B\mathbf{u}_i(s)\, ds, \qquad i = 1, 2,$$

and also

$$\mathbf{y} = e^{A(t_* - t_c)}\mathbf{z}_i + \int_{t_c}^{t_*} e^{A(t_* - s)} B\mathbf{u}_i(s)\, ds, \qquad i = 1, 2.$$

Therefore

$$\begin{aligned}\mathbf{p} &= e^{At_*}\mathbf{x}_0 + \int_0^{t_*} e^{A(t_*-s)}B\mathbf{v}(s)\,ds \\ &= \mathbf{y} - \int_{t_c}^{t_*} e^{A(t_*-s)}B\mathbf{u}_1(s)\,ds + \int_{t_c}^{t_*} e^{A(t_*-s)}B\mathbf{u}_2(s)\,ds \\ &= e^{A(t_*-t_c)}\mathbf{z}_1 + \int_{t_c}^{t_*} e^{A(t_*-s)}B\mathbf{u}_2(s)\,ds = e^{A(t_*-t_c)}\mathbf{z}_1 + \mathbf{y} - e^{A(t_*-t_c)}\mathbf{z}_2 \\ &= \mathbf{y} + e^{A(t_*-t_c)}(\mathbf{z}_1 - \mathbf{z}_2).\end{aligned}$$

Similarly, $\mathbf{q} = \mathbf{y} + e^{A(t_*-t_c)}(\mathbf{z}_2 - \mathbf{z}_1)$, and so $\mathbf{p} \neq \mathbf{q}$, with $\mathbf{y} = \frac{1}{2}(\mathbf{p}+\mathbf{q})$.

Conversely, suppose that $\mathbf{y}$ is not an extreme point of $K(t_*; \mathbf{x}_0)$, so $\mathbf{y} = \frac{1}{2}[\mathbf{y}_1 + \mathbf{y}_2]$ where $\mathbf{y}_1$ and $\mathbf{y}_2$ belong to $K(t_*; \mathbf{x}_0)$. By the Bang–Bang Principle there are bang–bang controls $\mathbf{u}_1(\cdot)$, $\mathbf{u}_2(\cdot)$ steering $\mathbf{x}_0$ to $\mathbf{y}_1$ and $\mathbf{y}_2$ respectively. Then $\mathbf{u}(t) = \frac{1}{2}[\mathbf{u}_1(t) + \mathbf{u}_2(t)]$ steers $\mathbf{x}_0$ to $\mathbf{y}$ and is *not* bang–bang. But the Bang–Bang Principle insists that there *is* a bang–bang control steering $\mathbf{x}_0$ to $\mathbf{y}$. Thus there is more than one control steering $\mathbf{x}_0$ to $\mathbf{y}$. By the preceding lemma, the response from $\mathbf{x}_0$ to $\mathbf{y}$ is not unique as well. □

Recall that a set is strictly convex if the boundary consists only of extreme points; equivalently, if two points belong to the set, then the open segment between them lies in the interior of the set. (See the Mathematical Appendix.)

Theorem 6 (A Geometric Characterization of Normality). *(LA) is normal on* $[0, t_*] \Leftrightarrow K(t_*; \mathbf{x}_0)$ *is strictly convex for one (hence all)* $\mathbf{x}_0$.

Proof. Suppose that (LA) is normal and that $\mathbf{y} \in \partial K(t_*; \mathbf{x}_0)$ (t_* may not be the optimal time for $\mathbf{y}$). We want to show that $\mathbf{y}$ is an extreme point of $\partial K(t_*; \mathbf{x}_0)$. The second part of the proof of Theorem 2 shows that all responses from $\mathbf{x}_0$ to $\mathbf{y}$ at time t_* must be extremal (lie on $\partial K(t; \mathbf{x}_0)$ for all $0 \leq t \leq t_*$) and Theorem 3 states that the associated controls must all be bang–bang (here is where we use normality). But if there were two bang–bang controls steering from $\mathbf{x}_0$ to $\mathbf{y}$ at time t_*, their average – which is *not* bang–bang – would also steer from $\mathbf{x}_0$ to $\mathbf{y}$ at time t_*. Thus there is exactly one control steering $\mathbf{x}_0$ to $\mathbf{y}$ (and it is bang–bang). By Theorem 5, this implies that $\mathbf{y}$ is an extreme point.

Conversely, let $K(t_*; \mathbf{x}_0)$ be strictly convex. Suppose that (LA) is not normal. Then there is a nonzero vector $\mathbf{h}$ such that some component, say the jth, of the row vector $\mathbf{h}^T e^{-At}B$ vanishes identically on $[0, t_*]$. Then $\mathbf{h}^T e^{-At}\mathbf{b}_j = 0$ where $\mathbf{b}_j$ is as usual the jth column of B.

Define the (column) vector $\mathbf{n}$ by $\mathbf{n}^T = \mathbf{h}^T e^{-At_*}$, and choose $\mathbf{q} \in \partial K(t_*; \mathbf{x}_0)$ such that $\mathbf{n}$ is normal to the support hyperplane through $\mathbf{q}$ (this can be done by starting with an arbitrary plane P through $K(t_*; \mathbf{x}_0)$ normal to $\mathbf{n}$, and translating P parallel to $\mathbf{n}$ until the last point $\mathbf{q}$ of contact with the strictly convex set $K(t_*; \mathbf{x}_0)$). Let $\mathbf{u}(\cdot) \in \mathcal{U}_m$ be chosen so that $\mathbf{x}(t_*; \mathbf{x}_0, \mathbf{u}(\cdot)) = \mathbf{q}$. By the proof of Theorem 3,

$$u^i(t) = \operatorname{sgn} \mathbf{h}^T e^{-At} \mathbf{b}_i, \qquad i = 1, \ldots, m.$$

Let $\mathbf{v}(\cdot) \in \mathcal{U}_m$ be defined by $v^i(t) \equiv u^i(t)$ for $i \neq j$, and $v^j(t) \neq u^j(t)$ on all of $[0, t_*]$. Since $\mathbf{q}$ is an extreme point of $K(t_*; \mathbf{x}_0)$, Theorem 5 implies that the response from $\mathbf{x}_0$ to $\mathbf{q}$ is unique. Then by the lemma preceding Theorem 5, we can conclude that the control steering $\mathbf{x}_0$ to $\mathbf{q}$ is unique. Therefore, since $\mathbf{u}(t) \not\equiv \mathbf{v}(t)$,

$$\mathbf{q} = \mathbf{x}(t_*; \mathbf{x}_0, \mathbf{u}(\cdot)) \neq \mathbf{x}(t_*; \mathbf{x}_0, \mathbf{v}(\cdot)).$$

But $\mathbf{n}^T\mathbf{q} = \mathbf{n}^T\mathbf{x}(t_*; \mathbf{x}_0, \mathbf{u}(\cdot)) = \mathbf{n}^T\mathbf{x}(t_*; \mathbf{x}_0, \mathbf{v}(\cdot))$, since by the response formula (4), the definition of $\mathbf{n}$, and the fact that $\mathbf{h}^T e^{-As}\mathbf{b}_j \equiv 0$,

$$\begin{aligned}
\mathbf{n}^T\mathbf{q} = \mathbf{n}^T\mathbf{x}(t_*; \mathbf{x}_0, \mathbf{u}(\cdot)) &= \mathbf{n}^T e^{At_*}\mathbf{x}_0 + \mathbf{n}^T \int_0^{t_*} e^{A(t_*-s)} B\mathbf{u}(s)\, ds \\
&= \mathbf{h}^T\mathbf{x}_0 + \int_0^{t_*} \mathbf{h}^T e^{-As} B\mathbf{u}(s)\, ds \\
&= \mathbf{h}^T\mathbf{x}_0 + \int_0^{t_*} \sum_{i=1}^m \mathbf{h}^T e^{-As}\mathbf{b}_i u^i(s)\, ds \\
&= \mathbf{h}^T\mathbf{x}_0 + \int_0^{t_*} \sum_{i=1}^m \mathbf{h}^T e^{-As}\mathbf{b}_i v^i(s)\, ds \\
&= \mathbf{n}^T\mathbf{x}(t_*; \mathbf{x}_0, \mathbf{v}(\cdot)).
\end{aligned}$$

Thus $\mathbf{n}^T[\mathbf{q} - \mathbf{x}(t_*; \mathbf{x}_0, \mathbf{v}(\cdot))] = 0$, so the line segment from $\mathbf{q}$ to $\mathbf{x}(t_*; \mathbf{x}_0, \mathbf{v}(\cdot))$ lies in the hyperplane P. But $K(t_*; \mathbf{x}_0)$ is convex, so each point on this segment is also in $K(t_*; \mathbf{x}_0)$, thus $K(t_*; \mathbf{x}_0)$ is not strictly convex – a contradiction. □

We can now give an analytic characterization of normality.

Theorem 7. *(LA) is normal on* $[0, t_*]$ *if and only if* $\{\mathbf{b}_j, A\mathbf{b}_j, \ldots, A^{n-1}\mathbf{b}_j\}$ *is a linearly independent set of vectors in* R^n *for each column vector* $\mathbf{b}_j$ *of* B, $j = 1, 2, \ldots, m$.

Proof. Suppose that for some j, $\{\mathbf{b}_j, A\mathbf{b}_j, \ldots, A^{n-1}\mathbf{b}_j\}$ is dependent. Then there is a vector $\mathbf{d} \neq 0$, such that

$$\mathbf{d}^T\mathbf{b}_j = \mathbf{d}^T A\mathbf{b}_j = \cdots = \mathbf{d}^T A^{n-1}\mathbf{b}_j = 0.$$

Now the $n \times n$ matrix A satisfies its own characteristic equation so we can write

$$A^n\mathbf{b}_j = \alpha_0\mathbf{b}_j + \alpha_1 A\mathbf{b}_j + \cdots + \alpha_{n-1}A^{n-1}\mathbf{b}_j, \tag{6}$$

and therefore $\mathbf{d}^T A^n \mathbf{b}_j = 0$. Multiplying (6) successively by A, $A^2, \ldots,$ we conclude that $\mathbf{d}^T A^k \mathbf{b}_j = 0$ for all integers $k \geq 0$. This implies in turn that $\mathbf{d}^T e^{-At}\mathbf{b}_j \equiv 0$ on $[0, t_*]$. Thus (LA) is not normal.

Conversely, suppose there is a $\mathbf{d} \neq \mathbf{0}$ in R^n and a column $\mathbf{b}_j$ of B such that $\mathbf{d}^T e^{-At}\mathbf{b}_j \equiv 0$ on $[0, t_*]$. Setting $t = 0$ we get $\mathbf{d}^T\mathbf{b}_j = 0$. Differentiating and then setting $t = 0$, we get $\mathbf{d}^T A\mathbf{b}_j = 0$. Continuing, we conclude that $\mathbf{d}^T A^h \mathbf{b}_j = 0$ for $h = 0, 1, \ldots, n-1$. Therefore $\mathbf{d}$ is perpendicular to all of the vectors $\{\mathbf{b}_j, A\mathbf{b}_j, \ldots, A^{n-1}\mathbf{b}_j\}$ and this set is linearly dependent. □

The above proof is very similar to the proof of Theorem 3 in Chapter II, in which we proved: "rank $M = n$ if and only if $\mathbf{0} \in \operatorname{Int} \mathscr{C}$," where $M = \{B, AB, \ldots, A^{n-1}B\}$ is the controllability matrix. There is in fact an obvious connection between the controllability matrix and the criterion for normality contained in Theorem 7:

$$\text{(LA) } \textit{normal} \Rightarrow \operatorname{rank} M = n.$$

Then, by Theorem 5 of Chapter II:

$$\text{(LA) } \textit{normal and } \operatorname{Re}\lambda \leq 0 \quad \textit{for all eigenvalues } \lambda \textit{ of } A \Rightarrow \mathscr{C} = R^n.$$

EXAMPLE 6 (Rank $M = n$ Need Not Imply Normality). Let $m = n = 2$, and $\dot{\mathbf{x}} = \mathbf{u}$. Then rank $M = \operatorname{rank}\{B, AB\} = 2$. But $\{\mathbf{b}_j, A\mathbf{b}_j\}$ is

$$\left\{\begin{bmatrix}1\\0\end{bmatrix}, \begin{bmatrix}0\\0\end{bmatrix}\right\} \quad (j=1), \quad \text{or} \quad \left\{\begin{bmatrix}0\\1\end{bmatrix}, \begin{bmatrix}0\\0\end{bmatrix}\right\} \quad (j=2),$$

therefore the system is not normal. Geometrically the general situation is as follows when $K(t; \mathbf{x}_0)$ is not a single point:

$$\begin{array}{ccc} \operatorname{rank} M = n & \Leftrightarrow & K(t; \mathbf{x}_0) \textit{ has nonempty interior} \\ \Uparrow & & \Uparrow \\ \text{(LA) } \textit{normal} & \Leftrightarrow & K(t; \mathbf{x}_0) \textit{ is strictly convex.} \end{array}$$

Corollary. *Let (LA) be normal. Then there is a neighborhood N of the origin such that every point of N can be steered to* $\mathbf{0}$ *by a unique time-optimal control which is bang–bang and piecewise constant. If* $\operatorname{Re}(\lambda) \leq 0$ *for each eigenvalue* λ *of A, then* $N = R^n$.

The proof immediately follows from the preceding theorem and Theorem 5 of Chapter II.

4. Applications

EXAMPLE 7 (The Rocket Car (Continuing Example 1)). We now apply the preceding results to the rocket car:

$$(R)\quad \dot{\mathbf{x}} = \begin{bmatrix} 0 & 1 \\ 0 & 0 \end{bmatrix}\mathbf{x} + \begin{bmatrix} 0 \\ 1 \end{bmatrix}u, \qquad B = \begin{bmatrix} 0 \\ 1 \end{bmatrix}, \qquad AB = \begin{bmatrix} 1 \\ 0 \end{bmatrix} \qquad (n = 2, m = 1).$$

It is easy to see that this is a normal system, and the only eigenvalue of A is $\lambda = 0$. The above corollary then applies. From the analysis in Example 1 of this chapter, we know that a time-optimal control must satisfy $\operatorname{sgn} u(t) = \operatorname{sgn}(\alpha - \beta t)$, so the time-optimal control has either no switch or one switch (Figure 7), exactly as our intuitive analysis showed in Chapter I.

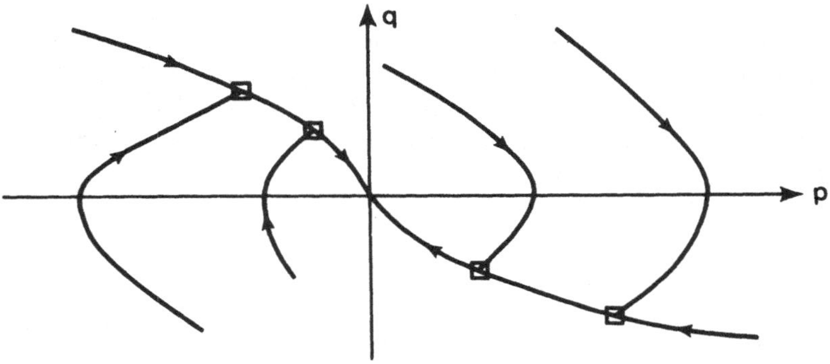

Figure 7

EXAMPLE 8 (The Linearized Pendulum (Continuing Example 2)). We have $(\omega(t) = \dot{\theta}(t))$:

$$\mathbf{x} = \begin{bmatrix} \theta(t) \\ \omega(t) \end{bmatrix}, \qquad \dot{\mathbf{x}} = \begin{bmatrix} 0 & 1 \\ -1 & 0 \end{bmatrix}\mathbf{x} + \begin{bmatrix} 0 \\ 1 \end{bmatrix}u, \qquad (m = 1, n = 2).$$

The set $\{\mathbf{b}_1, A\mathbf{b}_1\}$ is linearly independent, so the problem is normal, and the eigenvalues of A are $\pm i$, so the above corollary applies. Thus for each initial state $\mathbf{x}_0 \in R^2$, there exists a unique time-optimal control, which is bang–bang, steering $\mathbf{x}_0$ to $(0, 0)$. We showed in Example 2 that any time-optimal control satisfies $u(t) = \operatorname{sgn}\sin(t + \delta)$, so $u(t)$ changes sign with period π. To give a complete synthesis for this problem, we find the responses to $u(t) \equiv +1$ and $u(t) \equiv -1$. In the case $u(t) \equiv +1$ the system becomes

$$\dot{\theta} = \omega, \qquad \dot{\omega} = -\theta + 1.$$

Letting $\nu(t) = \theta(t) - 1$, we have $\dot{\nu} = \omega$, $\dot{\omega} = -\nu$ which represents clockwise motion around a circle centered at $\omega = \nu = 0$. In the original state space, we have clockwise motion around a circle centered at $\theta = +1$, $\omega = 0$, and each response traverses its circle in 2π units of time (Figure 8a). If $u(t) \equiv -1$, we can use a similar argument to show that the responses move clockwise around a circle centered at $\theta = -1$, $\omega = 0$ (Figure 8b), traversing the circle in time 2π.

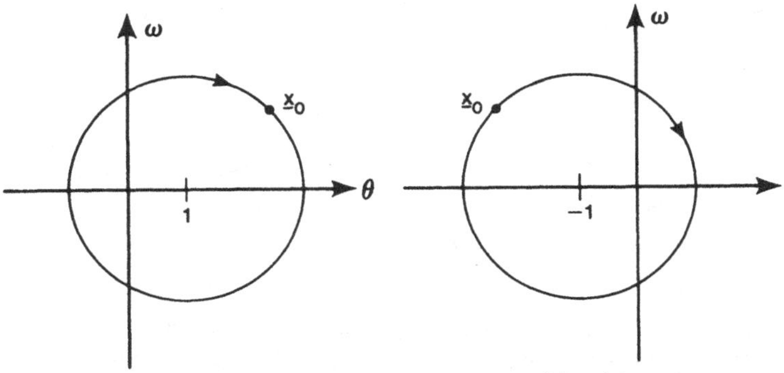

(a) $u(t) \equiv +1$. (b) $u(t) \equiv -1$

Figure 8

There is only one trajectory in each case which hits the target **0** (Figures 9a, 9b).

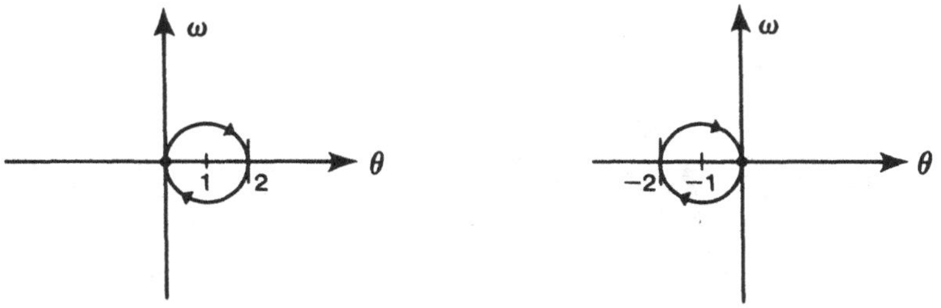

(a) Successful Response for $u(t) \equiv +1$ (b) Successful Response for $u(t) \equiv -1$

Figure 9

Clearly, we must get onto one of these to reach the target with a bang–bang control. Now $u(t)$ must change sign every π units of time, and π units of time is required to traverse a semicircle. Therefore, if we use the response in Figure 9a, for example, we must hit the lower semicircle in order to reach the origin (the upper semicircle is too far away from **0** in time). To hit this semicircle, we must have traversed a semicircle centered on $\theta = -1$, $\omega = 0$ (Figure 10a).

The easiest way to get an exact semicircle, ending at the switching state Δ, is to draw in the dashed semicircle and start the trajectory $(-)$ at the obvious switching point $\square$. The time to get from state $\square$ to state Δ is π, so we must have switched from a $u(t) \equiv +1$ response at $\square$ (Figure 10b), that is, from a circle centered on $\theta = 1$, $\omega = 0$. In order to get exact

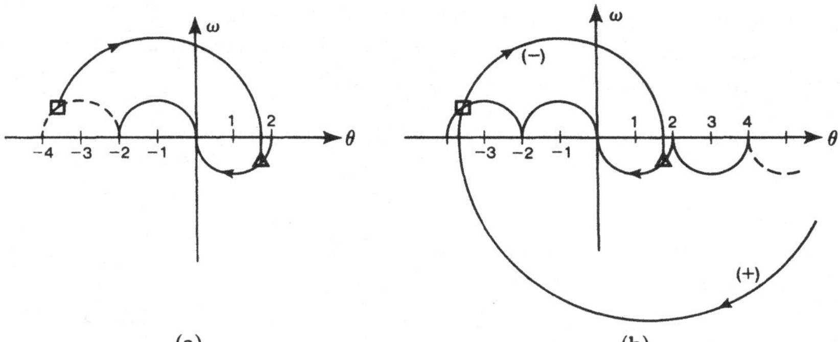

Figure 10

semicircles, we have sketched in "switching semicircles" on the positive x-axis. Continuing this analysis, we obtain a complete synthesis for the linearized pendulum (Figure 11). We switch control values whenever our response intersects the curve formed by the semicircles. Below this curve we use $u(t) \equiv +1$, above it we use $u(t) \equiv -1$. One response is drawn in.

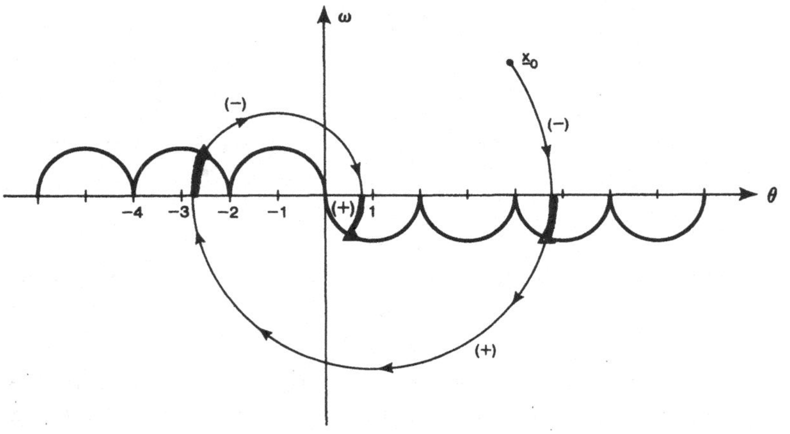

Figure 11

Notice that on three parts of the trajectory in Figure 11 (the darker segments) the control does not oppose the motion of the pendulum (u and $\dot{\theta} = \omega$ have the same sign). In contrast to the rocket car, the number of switches for the pendulum depends on how close the initial state is to the target $\mathbf{0}$ – the farther away we start, the more switches. The following theorem gives us some general information on the number of switches in one special case.

Theorem 8. *Suppose that (LA) is normal and that every eigenvalue of A is real. Then each component of any time-optimal control has at most $(n-1)$ switches.*

Proof. Since $\operatorname{sgn} u^i(t) = \operatorname{sgn} \mathbf{h}^T e^{-At}\mathbf{b}_i$, $i = 1, 2, \ldots, m$, we need only show that $\mathbf{h}^T e^{-At}\mathbf{b}_i$ has at most $(n-1)$ sign changes. We assume that the eigenvalues $\lambda_1, \ldots, \lambda_n$ of A are distinct. This is no restriction, since if not, we could perturb the entries of A by an arbitrarily small amount (preserving the number of sign changes for $\mathbf{h}^T e^{-At}\mathbf{b}_i$) to obtain a matrix $\tilde{A}$ with distinct eigenvalues. Now since $e^{-At}\mathbf{b}_i$ solves $\dot{\mathbf{x}} = -A\mathbf{x}$, we have $e^{-At}\mathbf{b}_i = \sum_{k=1}^n \mathbf{c}_k e^{-\lambda_k t}$, and

$$\mathbf{h}^T e^{-At}\mathbf{b}_i = \alpha_1 e^{-\lambda_1 t} + \alpha_2 e^{-\lambda_2 t} + \cdots + \alpha_n e^{-\lambda_n t} \equiv f_n(t).$$

Since (LA) is normal, $f_n(t)$ is nontrivial – at least one α_j is non-zero. We will show by induction on n that any such function can vanish at most $n-1$ times on R. For $n = 1$ this is clear. Suppose it is true for *any* such function $f_{n-1}(t)$. If there were a function $f_n(t)$ with n (or more) zeros on R, then $g_n(t) \equiv e^{\lambda_n t} f_n(t) = \alpha_1 e^{(\lambda_n - \lambda_1)t} + \cdots + \alpha_n$ would also have n zeros. Therefore $f_{n-1}(t) \equiv g_n'(t)$ would have $(n-1)$ zeros, a contradiction. □

5. The Converse of the Maximum Principle

The following theorem provides us with a partial converse to the maximum principle (corollary to Theorem 3). If (LA) is normal, then the hypothesis of Theorem 9 (that (LA) is proper) holds.

Theorem 9. *Suppose that (LA) is proper, that is* $\operatorname{rank} M = n$ *where M is the control matrix* $\{B, AB, \ldots, A^{n-1}B\}$. *Then any successful control* $\mathbf{u}(\cdot)$ *which satisfies the maximum principle: for some* $\mathbf{h} \neq \mathbf{0}$: $\mathbf{h}^T e^{-At}B\mathbf{u}(t) = \sup_{\mathbf{v}\in\Omega} \mathbf{h}^T e^{-At}B\mathbf{v}$, $0 \le t \le t_*$, *is time-optimal on* $[0, t_*]$.

Proof. Suppose that $\mathbf{u}_*(\cdot)$ satisfies the maximum principle for some $\mathbf{h} \in R^n$, on $[0, t_*]$ and assume that $\mathbf{x}_*[t] \equiv \mathbf{x}(t; \mathbf{x}_0, \mathbf{u}_*(\cdot))$ satisfies $\mathbf{x}_*[t_*] = \mathbf{0}$. Assume that there is a control $\mathbf{u}_\#(\cdot)$ which steers $\mathbf{x}_0$ to $\mathbf{0}$ at time $t_\# < t_*$ via the response $\mathbf{x}_\#[\cdot]$. We wish to derive a contradiction. By the response formula,

$$\begin{aligned} 0 &= \mathbf{h}^T e^{-At_\#}\mathbf{x}_\#[t_\#] = \mathbf{h}^T\mathbf{x}_0 + \int_0^{t_\#} \mathbf{h}^T e^{-As} B\mathbf{u}_\#(s)\, ds \\ &\Rightarrow \mathbf{h}^T\mathbf{x}_0 = -\int_0^{t_\#} \mathbf{h}^T e^{-As} B\mathbf{u}_\#(s)\, ds. \end{aligned} \tag{7}$$

Using the response formula for $\mathbf{u}_*(\cdot)$, we get

$$\mathbf{h}^T e^{-At_\#}\mathbf{x}_*[t_\#] = \mathbf{h}^T\mathbf{x}_0 + \int_0^{t_\#} \mathbf{h}^T e^{-As}B\mathbf{u}_*(s)\,ds$$

$$= \int_0^{t_\#} [\mathbf{h}^T e^{-As}B\mathbf{u}_*(s) - \mathbf{h}^T e^{-As}B\mathbf{u}_\#(s)]\,ds \geq 0,$$

where the inequality follows from the fact that $\mathbf{u}_*(t)$ maximizes $\mathbf{h}^T e^{-As}B\mathbf{v}$ over all $\mathbf{v}\in\Omega$. Now an easy differentiation shows that

$$\frac{d}{dt}[\mathbf{h}^T e^{-At}\mathbf{x}_*[t]] = \mathbf{h}^T e^{-At}B\mathbf{u}_*(t).$$

Since $\mathbf{h}^T e^{-At_*}\mathbf{x}_*[t_*] = 0$ ($\mathbf{u}(\cdot)$ is successful at time t_*), we have

$$0 \leq \mathbf{h}^T e^{-At_\#}\mathbf{x}_*[t_\#] = \mathbf{h}^T e^{-At_\#}\mathbf{x}_*[t_\#] - \mathbf{h}^T e^{-At_*}\mathbf{x}_*[t_*]$$

(8)

$$= \int_{t_*}^{t_\#} \mathbf{h}^T e^{-As}B\mathbf{u}_*(s)\,ds.$$

But $\mathbf{u}_*(s)$ maximizes $\mathbf{h}^T e^{-As}B\mathbf{v}$ for $\mathbf{v}\in\Omega$, and since $\mathbf{v}=\mathbf{0}$ is in Ω, this maximum is non-negative, i.e., $\mathbf{h}^T e^{-As}B\mathbf{u}_*(s) \geq 0$ on $[t_\#, t_*]$. Since $t_\# < t_*$, this means that the last integral in (8) is nonpositive, which implies that (8) collapses to equalities throughout, so

$$\int_{t_\#}^{t_*} \mathbf{h}^T e^{-As}B\mathbf{u}_*(s)\,ds = 0.$$

Since the integrand is non-negative, this in turn implies that $\mathbf{h}^T e^{-As}B\mathbf{u}_*(s) = 0$ a.e. on $[t_\#, t_*]$. Thus $\mathbf{h}^T e^{-As}B\mathbf{v} \leq 0$ for all $\mathbf{v}\in\Omega$ on $[t_\#, t_*]$, and if in fact $\mathbf{h}^T e^{-As}B\tilde{\mathbf{v}} < 0$ for some $\tilde{\mathbf{v}}\in\Omega$, then we would have $\mathbf{h}^T e^{-As}B(-\tilde{\mathbf{v}}) > 0$. Since $-\tilde{\mathbf{v}}\in\Omega$, this would be a contradiction. Therefore, $\mathbf{h}^T e^{-At}B\mathbf{v} = 0$ for all $\mathbf{v}\in\Omega$, $t_\# \leq t \leq t_*$. For each t and each $\mathbf{v}$ the expression $\phi(t) \equiv \mathbf{h}^T e^{-At}B\mathbf{v}$ is an analytic function of t, which must vanish identically since it vanishes on $[t_\#, t_*]$. Therefore $\phi(0) = \phi'(0) = \cdots = \phi^{(n-1)}(0) = 0$, which reduces to

$$\mathbf{h}^T B\mathbf{v} = \mathbf{h}^T AB\mathbf{v} = \cdots = \mathbf{h}^T A^{n-1}B\mathbf{v} = 0$$

for all $\mathbf{v}\in\Omega$. This implies that:

$$\mathbf{h}^T B = \mathbf{h}^T AB = \cdots = \mathbf{h}^T A^{n-1}B = \mathbf{0},$$

which implies that $\mathbf{h}$ is perpendicular to every column of M, contradicting our assumption that rank $M = n$. □

6. Extensions to More General Problems

First of all, we consider the system

$$\dot{\mathbf{x}} = A\mathbf{x} + B\mathbf{u} + \mathbf{c}(t), \tag{9}$$

where $\mathbf{c}(t)$ is a given continuous function from $[0, \infty)$ into R^n. The Bang–Bang Principle holds for this system, in fact as mentioned in Chapter II, it holds when A and B are continuous functions of t (Lee and Markus [1967], p. 164). The response formula for (9) is

$$\mathbf{x}[t] = e^{At}\left(\mathbf{x}_0 + \int_0^t e^{-As}\mathbf{c}(s)\,ds\right) + e^{At}\int_0^t e^{-As}B\mathbf{u}(s)\,ds.$$

For fixed t, introducing $\mathbf{c}(t)$ has just translated the response, so clearly the geometry and continuity of $K(t; \mathbf{x}_0)$ are unchanged. Also, any convex combination of two responses is again a response – in fact, to the same convex combination of the corresponding controls. Using these observations, one can simply parallel the proofs in the preceding sections to show that, *with the exception of the corollary to Theorem* 7, *every single result from this chapter holds for* (9).

For the general linear system

$$\dot{\mathbf{x}} = A(t)\mathbf{x} + B(t)\mathbf{u} + \mathbf{c}(t) \tag{L}$$

with continuous coefficients, the response formula is

$$\mathbf{x}[t] = X(t)\mathbf{x}_0 + \int_0^t X(t)X^{-1}(s)[B(s)\mathbf{u}(s) + \mathbf{c}(s)]\,ds,$$

where $X(t)$ is the fundamental matrix for $\dot{\mathbf{x}} = A(t)\mathbf{x}$ with $X(0) = I$. For an optimal control $\mathbf{u}(\cdot)$, the maximum principle can be written either as:

there exists a constant vector $\mathbf{h}_0$, *such that*

$$\mathbf{h}_0^T X^{-1}(t)B\mathbf{u}(t) = \max_{\mathbf{v}\in\Omega} \mathbf{h}_0^T X^{-1}(t)B\mathbf{v}, \tag{10}$$

or: *there exists* $\mathbf{h}(t)$ *such that*

$$\mathbf{h}^T(t)B\mathbf{u}(t) = \max_{\mathbf{v}\in\Omega} \mathbf{h}^T(t)B\mathbf{v} \tag{11}$$

where $\mathbf{h}(t)$ *is some solution of the adjoint system*

$$\dot{\mathbf{h}} = -A^T(t)\mathbf{h}. \tag{12}$$

These two formulations are equivalent, since $[X^T(t)]^{-1}$ is a fundamental matrix for (12), and therefore $\mathbf{h}(t) = [X^T(t)]^{-1}\mathbf{h}_0$ for some constant vector $\mathbf{h}_0$.

The system (L) is defined to be *normal* if for all $t > 0$, no two distinct (ignoring sets of measure zero) controls $\mathbf{u}_1(\cdot)$, $\mathbf{u}_2(\cdot)$ can steer an initial state $\mathbf{x}_0$ to the same state on $\partial K(t; \mathbf{x}_0)$. Normality is equivalent to the uniqueness of the function $\mathbf{u}(\cdot)$ satisfying (10) (or (11) for any given solution of (12)). Even with these modifications, not all of our theorems carry over; however, the propositions (a)–(e) stated at the beginning of this chapter do carry over to (L). For details, see Sections 2.2 and 2.5 of Lee and Markus [1967], and Sections V.7, V.8 of Berkowitz [1974].

Exercises

Remark: In these exercises, unless stated otherwise, $|u^i(t)| < 1$, $i = 1, 2, \ldots, m$.

1. Consider the one-dimensional system ($m = n = 1$) $\dot{x} = u$, $x_0 = -1$ with $\mathscr{T}(t) \equiv 0$. Show that there is an extremal control which is not optimal. (Hint: Since the state space is one-dimensional, you can sketch the reachable cone RC. The lower boundary of RC is a trajectory which never reaches the target.)

2. With $m = n = 2$, consider $\dot{\mathbf{x}} = \mathbf{u}$ with $\mathbf{x}_0 = (-1, 0)$, $\mathscr{T}(t) \equiv \mathbf{0}$. Sketch the reachable cone RC in (x^1, x^2, t)-space. Show that $K(t; \mathbf{x}_0)$ is always a square. Show that the control $u^1(t) \equiv 1$, $u^2(t) = \phi(t)$ with $\phi(\cdot)$ any function for which $\int_0^1 \phi(s)\, ds = 0$ is time optimal. Thus there are infinitely many bang–bang time optimal controls. Why does this not contradict Theorem 4 (compare also Theorems 5, 6, and 7)? Show that if $\mathbf{q}$ is a corner of $\partial K(t; \mathbf{x}_0)$ then the control steering to each $\mathbf{q}$ is unique, but the support hyperplane P and normal $\mathbf{h}$ are not unique. If $\mathbf{q}$ is not a corner point of $\partial K(t; \mathbf{x}_0)$, show that the support hyperplane and normal are unique, but the path to $\mathbf{q}$ is not unique.

3. Give a concise, precise proof that if $\lambda_1 > 0$, $\lambda_2 > 0$, $\lambda_1 + \lambda_2 = 1$, and $\mathbf{u}_1(\cdot)$, $\mathbf{u}_2(\cdot)$ are distinct bang–bang controls, then $\lambda_1 \mathbf{u}_1(\cdot) + \lambda_2 \mathbf{u}_2(\cdot)$ is not a bang–bang control.

4. Consider the system ($m = n = 2$)
$$\dot{\mathbf{x}}(t) = \begin{bmatrix} 1 & 0 \\ 1 & 0 \end{bmatrix} \mathbf{u}(t), \qquad \mathbf{x}_0 = \begin{bmatrix} -1 \\ -1 \end{bmatrix}, \qquad \mathscr{T}(t) \equiv \mathbf{0}.$$
Sketch RC (it is two-dimensional in (t, x^1, x^2)-space) and show that there are infinitely many support hyperplanes to any extreme point in $K(t; \mathbf{x}_0)$. Show that $t_1 = 1$ is optimal. Show that there is an infinite number of optimal controls, but the response is unique. Why is the lemma preceding Theorem 5 not violated?

5. (From the proof of the lemma preceding Theorem 5). Prove that if $\int_0^t e^{-As} B[\mathbf{w}(s) - \mathbf{u}(s)]\, ds = \mathbf{0}$ for $0 \le t \le t_*$, then $B[\mathbf{w}(t) - \mathbf{u}(t)] = \mathbf{0}$ a.e. on $[0, t_*]$. Here $\mathbf{w}(\cdot)$ and $\mathbf{u}(\cdot)$ are bounded measurable functions from $[0, t_*]$ into R^m.

6. Carry through the proof of Theorem 5 for the more general system $\dot{\mathbf{x}} = A\mathbf{x} + B\mathbf{u} + \mathbf{c}(t)$, assuming Theorems 1–4 and their corollaries are valid for this system.

7. (Exercise 6 from Chapter I). Give a complete synthesis for the time-optimal control problem ($m = n = 1$)
$$\dot{x}(t) = bx(t) + u(t), \qquad \mathscr{T}(t) \equiv 0, \qquad b < 0.$$

8. Refer to Section 5 of Chapter I and Example 1 of this chapter. Compute the optimal time for an initial point (p_0, q_0) above the switching curve $(+) \cup (-)$ (cf. especially Figure 10 of Chapter I).

9. Prove Theorems 1 and 2 for the case when $\mathcal{T}(t)$ is a compact continuously varying set.

Chapter IV

Existence Theorems for Optimal Control Problems

1. Introduction

In this chapter we discuss sufficient conditions for the existence of an optimal control for the general problem:

$$\text{(1)} \qquad \dot{\mathbf{x}} = \mathbf{f}(t, \mathbf{x}, \mathbf{u}), \qquad \mathbf{x}(0) = \mathbf{x}_0, \qquad \mathbf{u}(\cdot) \in \mathcal{U}_m,$$

with associated cost

$$\text{(2)} \qquad C[\mathbf{u}(\cdot)] = \int_0^{t_1} f^0(t, \mathbf{x}[t], \mathbf{u}(t))\, dt.$$

Here, $\mathbf{f}$, f^0 are given continuous functions with values in R^n and R, respectively, $\mathcal{T}(t) \equiv \mathbf{0}$, and t_1 (which depends on $\mathbf{u}(\cdot)$) is the time at which the response arrives at the target.

Our aim is to present a representative sample of theorems which assert that there exists at least one optimal control, that is, a control $\mathbf{u}_*(\cdot) \in \mathcal{U}_m$ for which $C[\mathbf{u}_*(\cdot)] \leq C[\mathbf{u}(\cdot)]$ for all $\mathbf{u}(\cdot) \in \mathcal{U}_m$. In this chapter we are interested in *sufficient* conditions for at least one optimal control to exist, while in Chapter V we will deal with *necessary* conditions which an optimal control must satisfy. To understand the distinction, consider the calculus problem of finding the minimum of a real-valued function $h(x, y)$ of two variables, on a set $\mathcal{G}$ in R^2. One *sufficient* condition for $h(\cdot, \cdot)$ to have a minimum is that $\mathcal{G}$ be compact and $h(\cdot, \cdot)$ be continuous (or at least lower semi-continuous). If $h(\cdot, \cdot)$ is twice continuously differentiable on $\mathcal{G}$, then a *necessary* condition for $h(\cdot, \cdot)$ to have a minimum at (x_*, y_*) inside $\mathcal{G}$ is: $h_x = h_y = 0$, $h_{xy}^2 - h_{xx}h_{yy} < 0$, $h_{xx} > 0$, at (x_*, y_*).

The reader should notice that the above sufficient condition for $h(\cdot, \cdot)$ to have a minimum, while reassuring, is not at all useful in helping us *find* the minimum. The necessary condition, on the other hand, gives a concrete method for searching – find the critical points ($h_x = h_y = 0$) and apply the second derivatives test at these points. The situation is similar in control theory. In this chapter, we present a reasonably broad survey of sufficient conditions for the existence of at least one optimal control, while Chapter V will be devoted to the most important necessary condition which an optimal control must satisfy – the Pontryagin Maximum Principle.

The level of mathematical abstraction in the present chapter is quite high, and the results will not help us find the optimal control(s). The reader with little taste for mathematical abstraction should perhaps first read Sections 1 and 2, followed by the statements of Theorems 1–4 (omitting proofs) and all of our examples.

An outline of the chapter is as follows: In Section 2 we describe the basic approach to existence proofs, and discuss several problems which can occur, thereby motivating certain basic assumptions. In Section 3 we state and prove a fairly straightforward existence theorem. In Section 4 we state and prove an existence theorem for the general system (1)(2) under a standard *convexity condition*, and describe generalizations. In Section 5 we state and prove an existence theorem for problems which are *linear in the state*. Finally, in Section 6 we apply our theorems to certain examples, including the rocket car.

2. Three Discouraging Examples. An Outline of the Basic Approach to Existence Proofs

Nonlinear differential equations can be very pathological. For example, the solution of the scalar initial-value problem

$$\frac{dy}{dt} = [y(t)]^2, \qquad y(0) = 1$$

is $y(t) = 1/(1-t)$, which is unbounded, and in fact *undefined* at $t = 1$. This equation (modulo a shift in coordinates) is the basis for the following example of a problem for which there is no optimal control because, roughly speaking, there is an unbounded sequence of responses which generates a decreasing sequence of costs.

EXAMPLE 1. Consider the problem ($n = 2, m = 1$)

$$\mathbf{x} = \begin{bmatrix} p \\ q \end{bmatrix}, \qquad \dot{p} = 1, \qquad \dot{q} = [q+1]^2 u, \qquad \mathbf{x}_0 = \begin{bmatrix} -2 \\ 0 \end{bmatrix}, \qquad \mathcal{T}(t) \equiv \mathbf{0},$$

$$C[u(\cdot)] = \int_0^{t_1} f^0(t, \mathbf{x}, u)\, dt,$$

where

$$f^0(t, \mathbf{x}, u) = \begin{cases} 1 & \text{if } q < 0; \\ 1/(q+1)^2 & \text{if } q \geq 0. \end{cases}$$

Notice that $t_1 = 2$ for any successful response, since $\dot{p} = 1$, $p[0] = -2$, and the target value of $p[t]$ is 0. For the extremal control $u(t) \equiv +1$, the (unsuccessful) response is $p[t] = t-2$, $q[t] = t/(1-t)$, which only exists for $0 \leq t < 1$. To generate an unbounded increasing sequence of successful responses we define, for any constant $0 < \alpha < 1$,

$$u_\alpha(t) = \begin{cases} \alpha & \text{for } 0 \leq t \leq 1, \\ -\alpha & \text{for } 1 < t \leq 2. \end{cases}$$

The corresponding successful response is

$$p_\alpha[t] = t-2, \qquad q_\alpha[t] = \begin{cases} \alpha t/(1-\alpha t), & 0 \leq t \leq 1; \\ \alpha(2-t)/[1-(2-t)\alpha], & 1 \leq t \leq 2. \end{cases}$$

As $\alpha \uparrow 1$, $q_\alpha[t]$ tends to a singular function (Figure 1). It is easy to see that $C[u_\alpha(\cdot)] \to \frac{2}{3}$ as $\alpha \uparrow 1$. Therefore, an optimal control $u_*(\cdot)$ would have to satisfy $C[u_*(\cdot)] \leq \frac{2}{3}$.

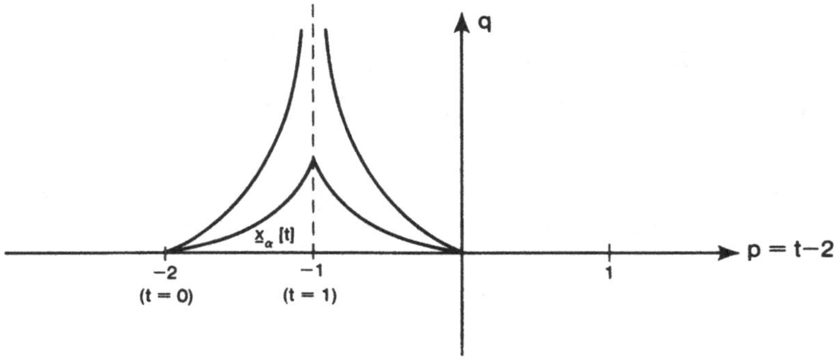

Figure 1

But we can show as follows that every admissible control generates a cost strictly greater than $\frac{2}{3}$. For a successful control, we must have $q[0] = q[2] = 0$, and the equation $\dot{q} = [q+1]^2 u$, with the restriction $-1 \leq u(t) \leq 1$, implies that

$$-[q+1]^2 \leq \dot{q}[t] \leq [q+1]^2.$$

The right-hand inequality implies that (i) $1 - t \leq 1/(q[t]+1)$ for $0 \leq t < 1$ (remember that $q[0] = 0$); the left inequality, integrated from t to 2 (with $t > 1$), implies that (ii) $t - 1 \leq 1/(q[t]+1)$ for $1 < t \leq 2$ (remember that $q[2] = 0$). But a successful response must be bounded, therefore we will have *strict* inequality in (i) for t in some interval. Thus,

$$C[u(\cdot)] > \int_0^1 (1-t)^2\,dt + \int_1^2 (t-1)^2\,dt = \tfrac{2}{3}.$$

Therefore our problem has no optimal control.

The preceding example was linear in the control $u(\cdot)$, while a nonlinearity in the state variable created difficulties. The next example shows that even when the responses are bounded and the problem is linear in the state, a nonlinearity in $u(\cdot)$ can lead to the nonexistence of an optimal control.

EXAMPLE 2. Consider the problem ($n=3$, $m=1$)

$$\mathbf{x} = \begin{bmatrix} p \\ q \\ r \end{bmatrix}, \qquad \dot{\mathbf{x}} = \begin{bmatrix} \sin 2\pi u \\ \cos 2\pi u \\ -1 \end{bmatrix}, \qquad \mathbf{x}_0 = \begin{bmatrix} 0 \\ 0 \\ 1 \end{bmatrix},$$

with $\mathcal{T}(t) = \mathbf{0}$, $C[u(\cdot)] = \int_0^1 \{[p[t]]^2 + [q[t]]^2\}\,dt$. The time of arrival at the target $\mathbf{0}$ is always 1, since $\dot{r} = -1$, $r[0] = 1$. Every response $\mathbf{x}[t]$ satisfies $|\dot{\mathbf{x}}[t]| \le 3$, so $\mathbf{x}[t]$ will satisfy the *a priori* bound $|\mathbf{x}[t]| \le |\mathbf{x}_0| + 3t \le 4$ for $0 \le t \le 1$. Suppose we could construct a sequence of controls $\{u_k(\cdot)\}$ for which $C[u_k(\cdot)] \to 0$. Since $C[u(\cdot)] \ge 0$ always, any optimal control $u_*(\cdot)$ would then have to give a cost $C[u_*(\cdot)] = 0$. This would lead to a contradiction, since $C[u_*(\cdot)] = 0$ implies $p_*[t] = q_*[t] = 0$ a.e., hence $\dot{p}_*[t] = \dot{q}_*[t] = 0$ a.e., which is impossible, since $\dot{p}_*[t] = \sin 2\pi u_*(t)$, $\dot{q}_*[t] = \cos 2\pi u_*(t)$.

The sequence of controls which will accomplish this is $u_k(t) = kt - [kt]$, $k = 1, 2, \ldots$, where $[\cdot]$ means "integer part" (Figure 2). Then $u_k(\cdot) \in \mathcal{U}_1[0, 1]$ and

$$\sin 2\pi u_k(t) = \sin 2\pi kt, \qquad \cos 2\pi u_k(t) = \cos 2\pi kt.$$

The associated responses are

$$p_k[t] = (1 - \cos 2\pi kt)/2\pi k, \qquad q_k[t] = \sin 2\pi kt/2\pi k, \qquad r_k[t] = 1 - t.$$

A direct evaluation shows that $C[u_k(\cdot)] = 1/2\pi^2k^2 \to 0$ as $k \to \infty$.

The preceding examples should convince the reader that some fairly restrictive hypotheses are needed in order to prove a general existence theorem for the problem (1) (2). To describe these, we introduce some notation. Let $\Delta(T)$ be the class of all admissible controls which steer $\mathbf{x}_0$ to the target $\mathbf{0}$ in time t_1, $0 < t_1 \le T$. One basic assumption will be that for some T, $\Delta(T)$ is non-empty, $\Delta(T) \ne \varnothing$, since you can't have an optimal control without at least one successful control. We will also assume that *successful responses on* $[0, T]$ *satisfy an a priori bound*:

(3) $$|\mathbf{x}(t; \mathbf{x}_0, \mathbf{u}(\cdot))| \le \alpha \quad \text{for all } \mathbf{u}(\cdot) \in \Delta(T), \quad 0 \le t \le t_1,$$

where $\alpha = \alpha(T)$ is a constant depending only on T. This condition is implied

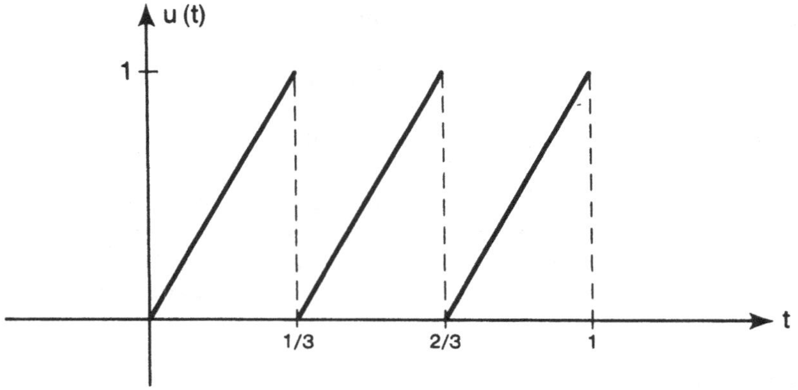

Figure 2 $u_3(t)$

by either of the following:

(a) $|\mathbf{f}(t, \mathbf{x}, \mathbf{u})| \leq \alpha|\mathbf{x}| + \beta, \qquad \left(|\mathbf{x}| = \sum_1^n |x^i|\right);$

(b) $|\mathbf{x}^T\mathbf{f}(t, \mathbf{x}, \mathbf{u})| \leq \alpha\|\mathbf{x}\|^2 + \beta, \qquad \left(\|\mathbf{x}\|^2 = \mathbf{x}^T\mathbf{x}\right);$

on $[0, T] \times R^n \times \Omega$, where α, β are nonnegative constants depending only on T. For example, to show that (a) implies (3), we write:

$$|\mathbf{x}[t]| = \left|\mathbf{x}_0 + \int_0^t \mathbf{f}(s, \mathbf{x}[s], \mathbf{u}(s))\, ds\right| \leq |\mathbf{x}_0| + \int_0^t \alpha|\mathbf{x}[s]|\, ds + \beta T.$$

Applying Gronwall's inequality we obtain:

$$|\mathbf{x}[t]| \leq (|\mathbf{x}_0| + \beta T)e^{\alpha T}.$$

Therefore *all* responses on $[0, T]$ (successful or not) satisfy an *a priori* bound. That (b) implies (3) is Exercise 1 at the end of this chapter. *Note that by the above argument, responses for the continuous linear system* $\dot{\mathbf{x}} = A(t, \mathbf{u})\mathbf{x} + \mathbf{b}(t, \mathbf{u})$ *from a fixed initial state satisfy an a priori bound.*

Even with the assumptions described above, Example 2 shows that we need some assumptions about dependence on the control. We will either restrict our control class or make a *convexity* assumption: the endpoints of the set of vectors

$$\hat{\mathbf{f}}(t, \mathbf{x}, \Omega) = \{(f^0(t, \mathbf{x}, \mathbf{v}), \mathbf{f}^T(t, \mathbf{x}, \mathbf{v}))^T | \mathbf{v} \in \Omega\}$$

form a convex set in R^{n+1} for each $(t, \mathbf{x})$.

Before we state and prove our existence theorems, we outline the general idea behind the proofs of such theorems. The problem (1) (2) is in essence a mapping

$$C: \mathbf{u}(\cdot) \to C[\mathbf{u}(\cdot)]$$

from $\mathcal{U}_m$ into the real numbers. This mapping can be extremely complicated since the cost functional $C[\mathbf{u}(\cdot)]$ usually involves the response $\mathbf{x}[\cdot]$. In principle, most existence theorems ought to be reducible (via the correct topology on $\mathscr{U}_m$) to the statement: "a lower semicontinuous real-valued function on a compact set attains its minimum on that set." However, it is often less complicated to directly approach an existence proof as follows:

(i) Show that $C[\mathbf{u}(\cdot)]$ is bounded below, hence there exists a minimizing sequence $\{\mathbf{u}_k(\cdot)\}$ with associated responses $\{\mathbf{x}_k[\cdot]\}$.
(ii) Show that $\{\mathbf{x}_k[\cdot]\}$ converges to a limit $\mathbf{x}_*[\cdot]$ (not necessarily a response).
(iii) Show that there is a $\mathbf{u}_*(\cdot) \in \mathscr{U}_m$ for which $\mathbf{x}_*[\cdot]$ is a response.

Notice that we make no claim about the convergence of $\{\mathbf{u}_k(\cdot)\}$ to $\mathbf{u}_*(\cdot)$. For problems linear in the control, we can first prove $\{\mathbf{u}_k(\cdot)\} \to \mathbf{u}_*(\cdot)$, and then prove $\mathbf{x}_k[\cdot] \to \mathbf{x}_*[\cdot]$ (cf. Strauss [1968], p. 81), but in general this is not possible. The following example shows that we can have a minimizing sequence of controls which does not converge in any sense, even though the associated responses converge to an optimal response!

EXAMPLE 3. Consider the problem ($n = 2$, $m = 2$)

$$\mathbf{x} = \begin{bmatrix} p \\ q \end{bmatrix}, \qquad \dot{p} = 1 - \|\mathbf{u}(t)\|^2, \qquad \dot{q} = 1,$$

$$\mathbf{x}_0 = \begin{bmatrix} 0 \\ 0 \end{bmatrix}, \qquad \mathscr{T}(t) = \begin{bmatrix} 0 \\ 1 \end{bmatrix}, \qquad C[\mathbf{u}(\cdot)] = \int_0^1 [p(t)]^2 \, dt.$$

Notice that $t_1 = 1$ since $\dot{q} = 1$ with $q(0) = 0$, $q(t_1) = 1$. Any control $\mathbf{u}(\cdot) \in \mathscr{U}_m[0, 1]$ with $\|\mathbf{u}(t)\|^2 \equiv 1$ is optimal, but notice that the optimal *response* $p(t) \equiv 0$, $q(t) = t$, is unique. Let

$$\mathbf{u}_k(t) = \begin{bmatrix} \sin kt \\ \cos kt \end{bmatrix}, \qquad k = 0, 1, 2, \ldots.$$

Then $\{\mathbf{u}_k(\cdot)\}$ is a minimizing sequence, but $\{\mathbf{u}_k(\cdot)\}$ does not converge in any usual sense to an optimal control. Obviously it does not converge pointwise. By the Riemann–Lebesgue Lemma,

$$\lim_{k \to \infty} \int_0^1 \phi(t) \sin kt \, dt = \lim_{k \to \infty} \int_0^1 \phi(t) \cos kt \, dt = 0,$$

for any $\phi(\cdot) \in L^2[0, 1]$. So this sequence converges weakly to $\mathbf{u}_*(t) \equiv \mathbf{0}$ in $L^2[0, 1]$, but $\mathbf{u}_*(t) \equiv \mathbf{0}$ is not successful. Notice that the associated responses $p_k[t] \equiv 0$, $q_k[t] = t$, do converge to an optimal response.

As if all of the above weren't enough, we also must be careful to keep the process restricted to a compact time interval. This is demonstrated by the minimum fuel problem for the rocket car (p. 15). In this example, the cost function was the amount of fuel used to move the car from one fixed point to another (with initial and terminal velocities zero). We saw that we could use an arbitrarily small amount of fuel by using an increasingly long time, and for this reason there was no optimal control.

3. Existence for Special Control Classes

As we mentioned above, existence theory is in essence a study of a continuous or lower semicontinuous function $C[\mathbf{u}(\cdot)]$ on a compact set $\mathcal{U}_m[0, T]$. For example, if $\mathcal{U}_m[0, T]$ is given the weak topology of $L^2[0, T]$, then $\mathcal{U}_m[0, T]$ is (sequentially) compact – this is the content of Proposition 1 in the Appendix to Chapter II. The requirements of the theorems of this chapter (e.g., $\hat{\mathbf{f}}(t, \mathbf{x}, \Omega)$ convex) are designed to imply that the function $C[\mathbf{u}(\cdot)]$ is lower semicontinuous in this weak topology.

If we restrict our controls to take their values in certain special subsets of $\mathcal{U}_m[0, T]$, then it is possible to choose these subsets so they are compact in stronger topologies, e.g., in the sup norm. Then we can weaken our restrictions on f^0 and $\mathbf{f}$ and still have $C[\mathbf{u}(\cdot)]$ lower semicontinuous. This is the motivation behind the Theorem 1 below, which deals with two special classes of controls, $\mathcal{U}_\lambda$ and $\mathcal{U}_r$. This first class is a compact subset of $\mathcal{U}_m$ in the sup norm, the second is compact in the L^1 norm.

For a given $\lambda > 0$, we define $\mathcal{U}_\lambda \subset \mathcal{U}_m$ to be those controls which satisfy a Lipschitz condition on their interval of definition:

$$|u(t) - u(s)| \leq \lambda |t - s|.$$

For a given integer $r \geq 0$, we define $\mathcal{U}_r \subset \mathcal{U}_m$ to be those controls which are piecewise constant with at most r points of discontinuity on their interval of definition.

Theorem 1. *Let* $[0, T]$ *be a fixed interval. Suppose that our usual control class* $\mathcal{U}_m[0, T]$ *is replaced by either* $\mathcal{U}_\lambda[0, T]$ *or* $\mathcal{U}_r[0, T]$ *for some fixed* $\lambda > 0$ *or integer* $r \geq 0$ *and assume* $\Delta(T) \neq \varnothing$. *Assume that* f^0 *and* $\mathbf{f}$ *are continuous, and that successful responses satisfy an a priori bound* (3). *Then there exists an optimal control.*

Proof. The first part of the argument applies to both $\mathcal{U}_\lambda$ and $\mathcal{U}_r$. Since $\|\mathbf{x}[t]\| \leq \alpha$ for any successful response, and f^0 is continuous on the compact set $[0, T] \times [-\alpha, \alpha] \times [-1, 1]$, it follows that $f^0(t, \mathbf{x}[t], \mathbf{u}(t))$ is uniformly bounded for all successful controls on $[0, T]$. Therefore, since we are restricted to $[0, T]$, $C[\mathbf{u}(\cdot)]$ is bounded below for $\mathbf{u}(\cdot) \in \Delta(T)$. Let $\{\mathbf{u}_k(\cdot)\}$ be a minimizing sequence for $C[\mathbf{u}(\cdot)]$ from the appropriate class, $\mathcal{U}_\lambda$ or $\mathcal{U}_r$, that is,

$$C[\mathbf{u}_k(\cdot)] \downarrow c = \inf C[\mathbf{u}(\cdot)], \quad \mathbf{u}_k(\cdot) \text{ defined on } [0, t_1(k)],$$

where the infimum is taken over the control class being used, $\mathcal{U}_\lambda$ or $\mathcal{U}_r$. Since the points $\{t_1(k)\}$ belong to $[0, T]$, we may assume that $t_1(k) \to t_1 \in [0, T]$. We will show that both $\{\mathbf{u}_k(\cdot)\}$ and the associated response sequence $\{\mathbf{x}_k[\cdot]\}$ are uniformly bounded and equicontinuous families on $[0, t_1]$, hence by passing to subsequences (again denoted $\{\mathbf{u}_k\}$, $\{\mathbf{x}_k\}$) we will have $\mathbf{u}_k(t) \to \mathbf{u}_*(t)$, $\mathbf{x}_k[t] \to \mathbf{x}_*[t]$ uniformly in t, for some continuous pair $\mathbf{u}_*(\cdot)$, $\mathbf{x}_*[\cdot]$.

But $\mathbf{x}_k[t] = \mathbf{x}_0 + \int_0^t \mathbf{f}(s, \mathbf{x}_k[s], u_k(s))\, ds$, so by taking the limit as $k \to \infty$, we will have

$$\mathbf{x}_*[t] = \mathbf{x}_0 + \int_0^t \mathbf{f}(s, \mathbf{x}_*[s], \mathbf{u}_*(s))\, ds, \qquad 0 \le t \le t_1.$$

Therefore, $\mathbf{x}_*[\cdot]$ will indeed be a response to $\mathbf{u}_*(\cdot)$, and $\mathbf{u}_*(\cdot)$ will be optimal, since $\{\mathbf{u}_k(\cdot)\}$ is minimizing $\mathbf{x}_k(\cdot) \to \mathbf{x}_*(\cdot)$ and $\mathbf{u}_k(\cdot) \to \mathbf{u}_*(\cdot)$ uniformly, and the cost function f^0 is continuous.

To complete the proof, then, we must prove that for both classes, $\mathscr{U}_\lambda$ and $\mathscr{U}_r$, the sequences $\{\mathbf{u}_k(\cdot)\}$, $\{\mathbf{x}_k[\cdot]\}$ are uniformly bounded and equicontinuous. First we have to make a technical comment. Whenever $t_1(k) < t_1$, we extend $\mathbf{u}_k(\cdot)$ to $[t_1(k), t_1]$ as the constant vector $\mathbf{u}_k(t_1(k))$, and we extend $\mathbf{x}_k[\cdot]$ as the constant $\mathbf{x}_k[t_1(k)]$. We cannot extend $\mathbf{x}_k[t]$ as a solution of $\dot{\mathbf{x}}_k = \mathbf{f}(t, \mathbf{x}_k, \mathbf{u}_k)$, because we might exceed our *a priori* bound.

Suppose we are dealing with the class $\mathscr{U}_\lambda[0, T]$. Then the *entire class* is equicontinuous and uniformly bounded ($\mathbf{u}(t) \in \Omega$), so certainly $\{\mathbf{u}_k(\cdot)\}$ is, also. The associated responses satisfy $\dot{\mathbf{x}}_k = \mathbf{f}(t, \mathbf{x}_k, \mathbf{u}_k)$, $\mathbf{x}(0) = \mathbf{x}_0$ on $[0, t_1(k)]$, $\dot{\mathbf{x}}_k = \mathbf{0}$ on $[t_1(k), t_1]$, and

$$(t, \mathbf{x}_k[t], \mathbf{u}_k(t)) \in [0, T] \times [-\alpha, \alpha] \times [-1, 1] \quad \text{on } [0, t_1].$$

Since $\mathbf{f}$ is continuous, it follows that $\{|\dot{\mathbf{x}}_k|\}$ is uniformly bounded, which implies that $\{\mathbf{x}_k[\cdot]\}$ is equicontinuous and uniformly bounded on $[0, t_1]$. This completes the proof for the class $\mathscr{U}_\lambda[0, T]$.

For the class $\mathscr{U}_r[0, T]$, we assume $r = 2$ for definiteness. Then for each $\mathbf{u}(\cdot) \in \mathscr{U}_r[0, t_1]$ we have

$$\mathbf{u}(t) = \begin{cases} \mathbf{a}, & 0 \le t < \sigma; \\ \mathbf{b}, & \sigma < t < \tau; \\ \mathbf{c}, & \tau < t \le t_1, \end{cases}$$

where $(\mathbf{a}, \mathbf{b}, \mathbf{c}, \sigma, \tau)$ depend on $\mathbf{u}(\cdot)$. Of course, we may have $\sigma = \tau$ (one jump), or $\sigma = \tau = t_1$ (no jumps). If $\{\mathbf{u}_k(\cdot)\}$ is a minimizing sequence, with $t_1(k) \to t_1$, then each $\mathbf{u}_k(\cdot)$ is described by a 5-tuple $(\mathbf{a}_k, \mathbf{b}_k, \mathbf{c}_k, \sigma_k, \tau_k)$. We may assume that each sequence converges: $\sigma_k \to \sigma_*$, $\tau_k \to \tau_*$, $\mathbf{a}_k \to \mathbf{a}_*$, $\mathbf{b}_k \to \mathbf{b}_*$, $\mathbf{c}_k \to \mathbf{c}_*$, since the vectors all belong to the unit cube in R^m, while σ_k, τ_k belong to $[0, t_1]$ (we ignore the behavior of $\mathbf{u}_k(\cdot)$ beyond t_1). We define $\mathbf{u}_*(t)$ on $[0, t_1]$ by $(\mathbf{a}_*, \mathbf{b}_*, \mathbf{c}_*, \sigma_*, \tau_*)$. In fact $\mathbf{u}_k(\cdot) \to \mathbf{u}_*(\cdot)$ *uniformly* on the set (Figure 3)

$$\mathscr{S}_\varepsilon = [0, \sigma_* - \varepsilon] \cup [\sigma_* + \varepsilon, \tau_* - \varepsilon] \cup [\tau_* + \varepsilon, t_1]$$

for $\varepsilon > 0$ small. (For $\sigma = \tau$, etc., the proof is even simpler.) The responses $\{\mathbf{x}_k[\cdot]\}$ to $\{\mathbf{u}_k(\cdot)\}$ are uniformly bounded and equicontinuous, by the same argument as for the class $\mathscr{U}_\lambda$ ($\mathbf{x}_k[0] = \mathbf{x}_0$, $|\dot{\mathbf{x}}_k|$ bounded), so a subsequence (again denoted $\{\mathbf{x}_k[\cdot]\}$) converges uniformly to a limit $\mathbf{x}_*(t)$ on $[0, t_1]$. If $\mathscr{S}_\varepsilon^c$ is the complement of $\mathscr{S}_\varepsilon$ in $[0, t_1]$, the fact that $\mathbf{f}(t, \mathbf{x}[t], \mathbf{u}(t))$ is continuous

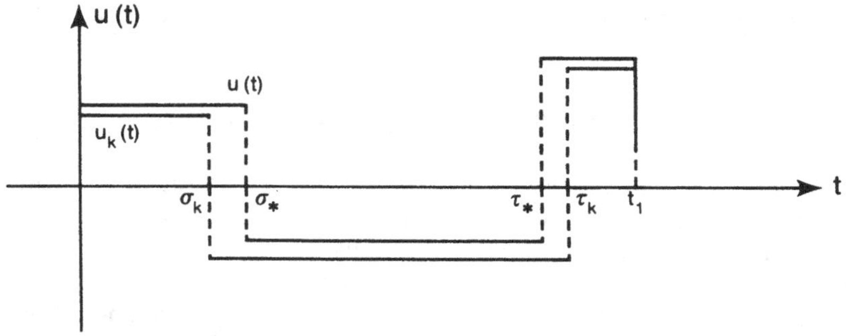

Figure 3 For k Large, $u_k(\cdot)$ Is Uniformly Close to $u_*(\cdot)$ Away from σ, τ.

and $(t, \mathbf{x}_k[t], \mathbf{u}_k(t))$ lies in a compact set implies that $\int_{\mathcal{S}_\varepsilon^c} |\mathbf{f}(t, \mathbf{x}_k[t], \mathbf{u}_k(t))|\, dt$ can be made arbitrarily small by choosing ε small.

Since $\mathbf{u}_k(\cdot) \to \mathbf{u}_*(\cdot)$ uniformly on $\mathcal{S}_\varepsilon$, and $\mathbf{x}_k[t] = \mathbf{x}_0 + \int_0^t \mathbf{f}(s, \mathbf{x}_k[s], \mathbf{u}_k(s))\, ds$, we can let $k \to \infty$, then $\varepsilon \to 0$, to conclude that

$$\mathbf{x}_*(t) = \mathbf{x}_0 + \int_0^t \mathbf{f}(s, \mathbf{x}_*[s], \mathbf{u}_*(s))\, ds.$$

Therefore, $\mathbf{x}_*[\cdot]$ is a response for $\mathbf{u}_*(\cdot)$. □

Remarks

1. The optimal response might hit the target $\mathbf{0}$ before the time t_1. It it does so, it must be at a higher or equal cost.
2. Theorem 1 remains valid for *fixed-endpoint* problems, i.e., both initial time t_0 and final time t_1 fixed in advance.
3. We can allow t_0 to vary within $[0, T]$.
4. We can drop the restriction to a fixed interval $[0, T]$, if we penalize responses that go for too long, to prevent t_0 and/or t_1 from becoming unbounded. For example, if $t_0 = 0$ is fixed and t_1 only restricted by $t_1 \geq 0$, then we need to assume $f^0(t, \mathbf{x}, \mathbf{v}) \geq \eta(t)$ for t large, with $\int^\infty \eta(t)\, dt = +\infty$. If t_0 varies, we need the restriction $f^0(t, \mathbf{x}, \mathbf{v}) \geq \eta(t)$ for t negative and large, with $\int_{-\infty} \eta(t)\, dt = +\infty$. (See Exercise 10 at the end of this chapter.)
5. The proof of Theorem 1 goes through with no essential changes if all successful *responses* are required to take values in some specified *compact* subset of R^n. This is sometimes called the "problem with restricted phase coordinates." For example, we could require an *a priori* bound plus

   $$h^i(x^1, x^2, \ldots, x^n) \leq 0, \qquad i = 1, \ldots, r$$

   for some specified continuous functions h^i. The important point is to keep $(t, \mathbf{x}(t), \mathbf{u}(t))$, $0 \leq t \leq t_1$, in a fixed compact set for all successful responses.
6. The proof of Theorem 1 goes through with no essential changes if the unit cube Ω – the range space of admissible controls – is replaced by a

continuously varying compact set $\Omega(t, \mathbf{x})$. (The convexity of Ω is irrelevant.)

7. One can add terms of the form $\phi(\mathbf{x}[t_1])$ and $\max_{[t_0,t_1]} \psi[\mathbf{x}(t)]$ to the cost, with ϕ, ψ continuous.
8. Both the initial state $\mathbf{x}_0$ and the target state $\mathcal{T}(t) \equiv \mathbf{0}$ may be replaced by continuously time-varying nonempty closed sets $X_0(t)$, $X_1(t)$ in R^n.

4. Existence Theorems under Convexity Assumptions

Theorem 2. *Consider the problem* (1) (2) *on a fixed interval* $[0, T]$, *with* $\mathbf{x}_0$ *given,* $\mathcal{T}(t) \equiv \mathbf{0}$ *and* $\mathbf{f}(t, \mathbf{x}, \mathbf{u})$, $f^0(t, \mathbf{x}, \mathbf{u})$ *continuous. Assume that* $\Delta(T) \neq \varnothing$, *and that successful responses satisfy an a priori bound* (3). *If, in addition, the set of points* $\hat{\mathbf{f}}(t, \mathbf{x}, \Omega) = \{(f^0(t, \mathbf{x}, \mathbf{v}), \mathbf{f}^T(t, \mathbf{x}, \mathbf{v}))^T | \mathbf{v} \in \Omega\}$ *is a convex set in* R^{n+1}, *then there exists an optimal control.*

Before we prove this theorem, we discuss the convexity hypothesis. The vector $\hat{\mathbf{f}}$ is often called the *extended velocity vector.*

EXAMPLE 4 (Examples of $\hat{\mathbf{f}}(t, \mathbf{x}, \Omega)$).

(a) Consider the scalar problem $\dot{x} = |u|^{1/2}$, $C[u(\cdot)] = \int_0^{t_1} |u(s)|^{1/2} x(s)\, ds$. Then $\hat{\mathbf{f}}(t, x, \Omega) = \{(|v|^{1/2}x, |v|^{1/2}) | -1 \leq v \leq 1\}$ is convex (Figure 4(a)).
(b) For the scalar system $\dot{x} = u$, $C[u(\cdot)] = \int_0^{t_1} [u(s)]^2\, ds$, $\hat{\mathbf{f}}(t, x, \Omega) = \{(v^2, v) | -1 \leq v \leq 1\}$ is *not* convex (Figure 4(b)) even though f is linear and f^0 is convex.
(c) If $\mathbf{f}(t, \mathbf{x}, \mathbf{u}) = A(t, \mathbf{x})\mathbf{u} + \mathbf{g}(t, \mathbf{x})$, $f^0(t, \mathbf{x}, \mathbf{u}) = \mathbf{a}(t, \mathbf{x})^T\mathbf{u} + g^0(t, \mathbf{x})$, with A an $n \times m$ matrix, $\mathbf{g}$ an n-vector, $\mathbf{a}$ an m-vector and g^0 real-valued, then $\hat{\mathbf{f}}(t, \mathbf{x}, \Omega) = \{(\mathbf{a}^T\mathbf{v} + g^0, A\mathbf{v} + \mathbf{g}) | \mathbf{v} \in \Omega\}$ is convex. Therefore, Theorem 1 covers the case when $\mathbf{f}$ and f^0 are both linear in the control.

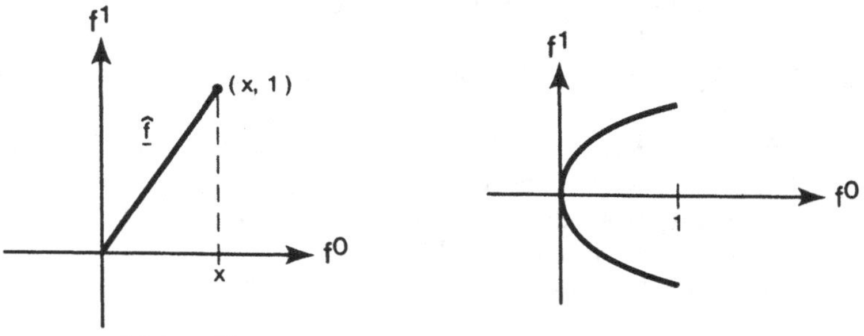

(a) $(x > 0)$ (b)

Figure 4 $\hat{\mathbf{f}}(t, \mathbf{x}, \Omega)$

The reader can see from Example 4 that the assumption "$\hat{\mathbf{f}}(t, \mathbf{x}, \Omega)$ is convex" is roughly a statement about the geometry of the relationship between f^0 and $\mathbf{f}$. This assumption does not imply that $\mathbf{f}$ and/or f_0 are convex functions of $\mathbf{u}$ (Example 4(a)), nor does the converse hold (Example 4(b)).

Proof. As we argued in the proof of Theorem 1, the cost functional $C[\mathbf{u}(\cdot)]$ is bounded below on the set $\Delta(T)$ of successful responses. Therefore, $c = \inf_\Delta C[\mathbf{u}(\cdot)]$ exists. Let $\{\mathbf{u}_k(\cdot)\}$ be a minimizing sequence for $C[\mathbf{u}(\cdot)]$:

$$C[\mathbf{u}_k(\cdot)] \equiv c_k \downarrow c.$$

For any control $\mathbf{u}(\cdot) \in \Delta(T)$ we define the extended response vector in R^{n+1}:

$$\hat{\mathbf{x}} = \begin{bmatrix} x^0 \\ \mathbf{x} \end{bmatrix}, \quad \text{where } \dot{x}^0 = f^0(t, \mathbf{x}, \mathbf{u}), \qquad x^0(0) = 0,$$

so that $x^0[t] = \int_0^t f^0(s, \mathbf{x}[s], \mathbf{u}(s))\, ds$, $x^0[t_1] = C[\mathbf{u}(\cdot)]$. Each pair $(\mathbf{u}_k(\cdot), \hat{\mathbf{x}}_k[\cdot])$ is defined on some interval $[0, t_1(k)]$, and we may assume that $t_1(k) \to t_1 \leq T$. Whenever $t_1(k) < t_1$ we extend $\hat{\mathbf{x}}_k[\cdot]$ to $[t_1(k), t_1]$ as the constant vector $\hat{\mathbf{x}}_k[t_1(k)]$.

The dynamics of our extended system is

$$\frac{d\hat{\mathbf{x}}}{dt} = \hat{\mathbf{f}}(t, \mathbf{x}, \mathbf{u}), \qquad \hat{\mathbf{x}}[0] = \begin{bmatrix} 0 \\ \mathbf{x}_0 \end{bmatrix}, \quad \text{where } \hat{\mathbf{f}} = \begin{bmatrix} f^0 \\ \mathbf{f} \end{bmatrix}.$$

We will show that there is a response $\hat{\mathbf{x}}_*[t]$ for this system which satisfies

$$\hat{\mathbf{x}}_*[t_1] = \begin{bmatrix} c \\ \mathbf{0} \end{bmatrix},$$

which means that the associated control $\mathbf{u}_*(\cdot)$ is optimal. The proof is carried out by means of the following propositions:

(i) The sequence $\{\hat{\mathbf{x}}_k[\cdot]\}$ is uniformly bounded and equicontinuous on $[0, t_1]$ so a subsequence (again denoted $\{\hat{\mathbf{x}}_k[\cdot]\}$) converges uniformly to a continuous limit $\hat{\mathbf{x}}_*(t)$. Since $\hat{\mathbf{x}}_k[t_1(k)] = (c_k, \mathbf{0})^T$, and $t_1(k) \to t_1$, we will have $\hat{\mathbf{x}}_*(t_1) = (c, \mathbf{0})^T$. Therefore, *if* $\hat{\mathbf{x}}_*[\cdot]$ is a response, then it is optimal.

(ii) The function $\hat{\mathbf{x}}_*[t]$ is absolutely continuous and $\int_{\mathscr{A}} (d\hat{\mathbf{x}}_k/dt) \to \int_{\mathscr{A}} (d\hat{\mathbf{x}}_*/dt)$ for any measurable subset $\mathscr{A}$ of $[0, t_1]$. (Equivalently: $d\hat{\mathbf{x}}_k/dt$ converges weakly in $L^2[0, t_1]$ to $d\hat{\mathbf{x}}_*/dt$, i.e., $\int_0^{t_1} \phi(t)\, d\hat{\mathbf{x}}_k/dt \to \int_0^{t_1} \phi(t)\, d\mathbf{x}_*/dt$ for all $\phi(\cdot) \in L^2[0, t_1]$.)

(iii) The function $\hat{\mathbf{x}}_*[t]$ satisfies the generalized differential equation

$$\frac{d\hat{\mathbf{x}}_*}{dt} \in \hat{\mathbf{f}}(t, x_*(t), \Omega) \quad \text{for } 0 \leq t \leq t_1.$$

(iv) (Fillipov's Lemma). There exists a control $\mathbf{u}_*(\cdot) \in \mathcal{U}_m[0, t_1]$ such that $d\hat{\mathbf{x}}_*/dt = \hat{\mathbf{f}}(t, \mathbf{x}_*, \mathbf{u}_*)$. Thus $\mathbf{u}_*(\cdot)$ is optimal, with response $\mathbf{x}_*[t]$.

Proof of (i). Since $\hat{\mathbf{f}}$ is continuous, $\{d\hat{\mathbf{x}}_k/dt\}$ is uniformly bounded on $[0, t_1]$, with $\hat{\mathbf{x}}_k[0] = (0, \mathbf{x}_0)^T$ so $\{\hat{\mathbf{x}}_k[\cdot]\}$ is uniformly bounded and equicontinuous. We select a subsequence (again denoted $\{\hat{\mathbf{x}}_k[\cdot]\}$) which converges uniformly on $[0, t_1]$ to a continuous limit $\hat{\mathbf{x}}_*(\cdot)$. As remarked above, $\hat{\mathbf{x}}_*(t_1) = (c, \mathbf{0})^T$.

Proof of (ii). Define $\hat{\mathbf{f}}_k[s] \equiv \hat{\mathbf{f}}_k(s, \mathbf{x}_k[s], \mathbf{u}_k(s))\, ds$. Then, since $\hat{\mathbf{x}}_k[t] = \hat{\mathbf{x}}_0 + \int_0^t \hat{\mathbf{f}}_k[s]\, ds$, we have $|\hat{\mathbf{x}}_k[t] - \hat{\mathbf{x}}_k[t']| \leq l|t - t'|$ where l is any constant satisfying $|\hat{\mathbf{f}}(t, \mathbf{x}, \mathbf{u})| \equiv |f^0(t, \mathbf{x}, \mathbf{u})| + |\mathbf{f}(t, \mathbf{x}, \mathbf{u})| \leq l$ on $[0, T] \times [-\alpha, \alpha] \times [-1, 1]$. Therefore $\hat{\mathbf{x}}_*(t)$ is absolutely continuous on $[0, t_1]$, so $d\hat{\mathbf{x}}_*/dt$ exists a.e., and as $k \to \infty$

$$\hat{\mathbf{x}}_k[t] = \hat{\mathbf{x}}_0 + \int_0^t \hat{\mathbf{f}}_k[s]\, ds \to \hat{\mathbf{x}}_*(t) = \hat{\mathbf{x}}_0 + \int_0^t (d\hat{\mathbf{x}}_*/ds)\, ds.$$

This means that $\int_0^t \hat{\mathbf{f}}_k[s]\, ds \to \int_0^t (d\hat{\mathbf{x}}_*/ds)\, ds$ for any $0 \leq t \leq t_1$, which in turn implies that $\int_I \hat{\mathbf{f}}_k \to \int_I d\hat{\mathbf{x}}_*/ds$ for any interval $I \subset [0, t_1]$. This implies that $\int_{\mathcal{A}} (d\hat{\mathbf{x}}_k/dt) \equiv \int_{\mathcal{A}} \hat{\mathbf{f}}_k \to \int_{\mathcal{A}} d\hat{\mathbf{x}}_*/ds$ for any measurable set $\mathcal{A} \subset [0, t_1]$.

Proof of (iii). Let $S = \{t | (d\hat{\mathbf{x}}_*/dt) \notin \hat{\mathbf{f}}(t, \mathbf{x}_*(t), \Omega)\}$. We want to show that the measure of S is 0, $|S| = 0$. Assume the contrary, $|S| > 0$. Since for each $t \in S$, the set $\hat{\mathbf{f}}(t, \mathbf{x}_*(t), \Omega)$ is *convex* and compact, there is a vector $\hat{\mathbf{b}}(t)$ and a number $\alpha(t)$ such that the hyperplane $P(t) = \{\hat{\mathbf{x}} | \langle \hat{\mathbf{b}}, \hat{\mathbf{x}} \rangle = \alpha\}$ separates $d\hat{\mathbf{x}}_*/dt$ from this set in R^{n+1} (Figure 5). Therefore (see the Mathematical Appendix) $\hat{\mathbf{b}}^T d\hat{\mathbf{x}}_*/dt > \alpha$, $\hat{\mathbf{b}}^T \hat{\mathbf{f}}(t, \mathbf{x}_*(t), \mathbf{v}) \leq \alpha$ for $\mathbf{v} \in \Omega$, so

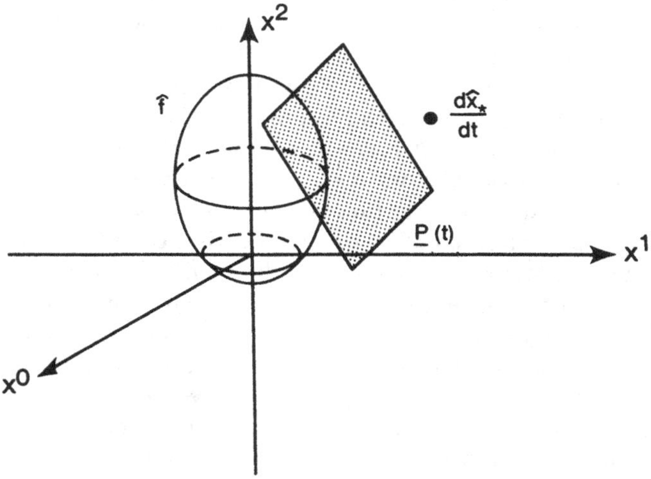

Figure 5 ($n = 2$)

$$
\begin{aligned}
\hat{\mathbf{b}}^T(t)(d\hat{\mathbf{x}}_*/dt) &> \max_{\mathbf{v}\in\Omega} \hat{\mathbf{b}}^T(t)\hat{\mathbf{f}}(t, \mathbf{x}_*(t), \mathbf{v}) \\
&\geq \limsup_{j\to\infty} \hat{\mathbf{b}}^T(t)\hat{\mathbf{f}}(t, \mathbf{x}_*(t), \mathbf{u}_j(t)) \qquad (4)\\
&= \limsup_{j\to\infty} \hat{\mathbf{b}}^T(t)\hat{\mathbf{f}}(t, \mathbf{x}_j(t), \mathbf{u}_j(t))
\end{aligned}
$$

(since $\hat{\mathbf{f}}$ is continuous and $\mathbf{x}_j[t]\to\mathbf{x}_*(t)$ uniformly). Because the inequality above is strict, we may assume $\hat{\mathbf{b}}(t)$ has rational entries. The set of all possible vectors $\hat{\mathbf{b}}(t)$ is thus denumerable, while $|S|>0$, therefore a *fixed* vector $\hat{\mathbf{b}}_0$ satisfies (4) in a set $\mathscr{A}\subset S$, $|\mathscr{A}|>0$. Then (remember $d\hat{\mathbf{x}}_j/dt = \hat{\mathbf{f}}(t, \mathbf{x}_j, \mathbf{u}_j)$)

$$
\int_{\mathscr{A}} \hat{\mathbf{b}}_0^T(d\hat{\mathbf{x}}_*/dt) > \int_{\mathscr{A}} \limsup_{j\to\infty} \hat{\mathbf{b}}_0^T\hat{\mathbf{f}}(t, \mathbf{x}_j, \mathbf{u}_j)\,dt > \limsup_{j\to\infty} \int_{\mathscr{A}} \hat{\mathbf{b}}_0^T(d\hat{\mathbf{x}}_j/dt),
$$

the last by Fatou's Lemma. But this contradicts the weak convergence of $d\hat{\mathbf{x}}_j/dt$ to $d\hat{\mathbf{x}}_*/dt$. Therefore, $(d\hat{\mathbf{x}}_*/dt)\in\hat{\mathbf{f}}(t, \mathbf{x}_*[t], \Omega)$ a.e., and so we can redefine $d\mathbf{x}_*/dt$ on a set of measure zero and drop the "a.e.".

Finally, to prove (iv) we state and prove:

Fillipov's Lemma. *Let $\Omega\subset R^m$ be compact, $\hat{\mathbf{g}}(t,\mathbf{u})$ a continuous function from $R\times\Omega$ into R^{n+1}. Suppose $\hat{\boldsymbol{\psi}}(\cdot)$ is a bounded measurable function from R into R^{n+1}, with $\hat{\boldsymbol{\psi}}(t)\in\hat{\mathbf{g}}(t,\Omega)$. Then there is a measurable $\mathbf{u}(\cdot)$ with $\mathbf{u}(t)\in\Omega$ for all t, such that $\hat{\boldsymbol{\psi}}(t)=\hat{\mathbf{g}}(t,\mathbf{u}(t))$.*

Proposition (iv) follows by taking $\hat{\boldsymbol{\psi}} = d\hat{\mathbf{x}}_*/dt$, $\hat{\mathbf{g}}(t,\mathbf{u}) = \hat{\mathbf{f}}(t, \mathbf{x}_*(t), \mathbf{u})$.

Proof of Fillipov's Lemma. For fixed $t\in[0, t_1]$, the set $\mathscr{E} = \{\mathbf{v}\in\Omega \mid \hat{\boldsymbol{\psi}}(t)=\hat{\mathbf{g}}(t,\mathbf{v})\}$ is nonempty. Let $\mathscr{E}_1$ be that subset of $\mathscr{E}$ for which v^1 is as small as possible. Ω is compact and $\hat{\mathbf{g}}$ is continuous, therefore $\mathscr{E}$ is compact and $\mathscr{E}_1$ is nonempty. Let $\mathscr{E}_2$ be that subset of $\mathscr{E}_1$ for which the second components v^2 are as small as possible. Continuing in this way, we obtain a nonempty set $\mathscr{E}_m$. A moment's thought shows that $\mathscr{E}_m$ is a single vector (compare any two vectors in $\mathscr{E}_m$), and we define $\mathbf{u}(t)$ to be this vector.

We claim that the resulting function $\mathbf{u}(\cdot)$ is measurable on $[0, t_1]$. We prove this by showing that $\mathscr{L}_\alpha = \{t \mid u^1(t)\leq\alpha\}$ is closed for any real α (an induction on the components completes the argument). First, we note that by Lusin's Theorem, for each integer $k>0$, $\hat{\boldsymbol{\psi}}(t)$ is continuous on a closed subset $\mathscr{F}_k$ of $[0, t_1]$ with $|[0, t_1]\backslash\mathscr{F}_k|<1/2^k$. We will show that $\mathscr{L}_\alpha\cap\mathscr{F}_k$ is closed for every α and any integer $k>0$. Suppose not. Then there is a sequence $\{t_j\}\subset\mathscr{L}_\alpha\cap\mathscr{F}_k$ converging to $\bar{t}\notin\mathscr{L}_\alpha\cap\mathscr{F}_k$, $u^1(t_j)\leq\alpha<u^1(\bar{t})$. Since $\mathscr{F}_k$ is closed, $\bar{t}\in\mathscr{F}_k$, and since Ω is compact, we can assume that the sequence of vectors $\{\mathbf{u}(t_j)\}$ converges to a limit $\mathbf{v}\in\Omega$. Then $u^1(t_j)\to v^1$,

$$
v^1\leq\alpha<u^1(\bar{t}).
$$

Now $\hat{\psi}(t_j) \to \hat{\psi}(\bar{t})$ because $\hat{\psi}(\cdot)$ is continuous on $\mathscr{F}_k$. Also, $\hat{\psi}(t_j) = \hat{\mathbf{g}}(t_j, \mathbf{u}(t_j))$, so

$$\hat{\psi}(\bar{t}) = \hat{\mathbf{g}}(\bar{t}, \mathbf{v}), \quad \text{with } v^1 \leq \alpha < u^1(\bar{t}).$$

This contradicts our algorithm for constructing $\mathbf{u}(\cdot)$, so we must conclude that $\mathscr{L}_\alpha \cap \mathscr{F}_k$ is closed for any α and any k.

Thus $\mathscr{L}_\alpha \cap \mathscr{F}_k$ is measurable, and so $\bigcup_{k=1}^{\infty} (\mathscr{L}_\alpha \cap \mathscr{F}_k) = \mathscr{L}_\alpha \cap (\bigcup_{k=1}^{\infty} \mathscr{F}_k)$ is measurable. But $\bigcup_{k=1}^{\infty} \mathscr{F}_k$ differs from $[0, t_1]$ by a set of measure zero so $\mathscr{L}_\alpha \cap (\bigcup_{k=1}^{\infty} \mathscr{F}_k)$ differs from $\mathscr{L}_\alpha \cap [0, t_1] = \mathscr{L}_\alpha$ by a measurable set, hence $\mathscr{L}_\alpha$ is also measurable. □

The remarks following the proof of Theorem 1 carry over verbatim to Theorem 2. In most cases, the proofs of the extensions are straightforward. For a full discussion, see Lee and Markus [1967], pp. 259–281.

The assumption in Theorem 2 that $\hat{\mathbf{f}}(t, \mathbf{x}, \Omega)$ is a convex subset of R^{n+1} implies that the sets $f^0(t, \mathbf{x}, \Omega)$ and $\mathbf{f}(t, \mathbf{x}, \Omega)$ are convex in R and R^n, respectively. This is easily seen using an argument by contradiction. The following corollary to Theorem 2 shows that whenever $\mathbf{f}(t, \mathbf{x}, \Omega)$ is convex, the reachable set $K(t; \mathbf{x}_0)$ is convex, compact, and continuous as a function of t. This result was proved for linear autonomous systems as the Lemma of Section 1, Chapter III and used extensively in our analysis of time-optimal problems. This corollary establishes the result, in particular, for the general linear system $\dot{\mathbf{x}} = A(t)\mathbf{x} + B(t)\mathbf{u}$, and we can then extend many of the theorems of Chapter III to this system.

Corollary. *For problems* (1) (2), *under the assumptions of Theorem* 2, *the reachable set* $K(t; \mathbf{x}_0)$ *is compact, and varies continuously with time.*

Proof. The convexity immediately follows from the response formula $\mathbf{x}[t] = \mathbf{x}_0 + \int_0^t \mathbf{f}(s, \mathbf{x}[s], \mathbf{u}(s))\, ds$ and the convexity of $\mathbf{f}(s, \mathbf{x}[s], \Omega)$.

Since responses satisfy an *a priori* bound, the set $K(t, \mathbf{x}_0)$ is bounded. To see that it is closed, hence compact, let $\{\mathbf{x}_k\} \subset K(t, \mathbf{x}_0)$ with $\mathbf{x}_k \to \mathbf{x}_* \in R^n$. Then each $\mathbf{x}_k$ is the value of a response $\mathbf{x}_k[s]$ at $s = t$. We now repeat the proofs of propositions (i)–(iv) from the proof of the theorem *word for word, with the "hats" removed* – that is, working with $\mathbf{x}[t]$, $\mathbf{f}(t, \mathbf{x}, \mathbf{u})$ rather than $\hat{\mathbf{x}}[t]$, $\hat{\mathbf{f}}(t, \mathbf{x}, \mathbf{u})$. The conclusion is that the sequence of responses $\{\mathbf{x}_k[s]\}$ converges to a function $\mathbf{x}_*(s)$ with $\mathbf{x}_*(t) = \mathbf{x}_*$, and $\mathbf{x}_*(s)$ is indeed a response for some admissible control, i.e., $\mathbf{x}_* \in K(t; \mathbf{x}_0)$.

To prove that $K(t; \mathbf{x}_0)$ is continuous, we must show that $K(t; \mathbf{x}_0)$ is contained in an ε-sack about $K(t_*; \mathbf{x}_0)$, and vice-versa, for $|t - t_*|$ small. Now for a given control we have

$$|\mathbf{x}[t] - \mathbf{x}[t_*]| \leq \int_t^{t_*} |\mathbf{f}(r, \mathbf{x}[r], u(r))|\, dr < \varepsilon \quad \text{if } |t - t_*| < \delta(\varepsilon).$$

(If the control only exists on the shorter of the intervals $[0, t]$, $[0, t_*]$, we can extend it as $\mathbf{0}$ and still have $|\mathbf{x}[t] - \mathbf{x}[t_*]| < \varepsilon$ for $|t - t_*|$ small.) Therefore

$K(t; \mathbf{x}_0)$ is contained in an ε-sack about $K(t_*; \mathbf{x}_0)$ for any t_* close enough to t, and vice-versa. □

Finally, we state without proof an existence theorem far more general than Theorem 2. The difficulty of the proof makes it beyond the level of this monograph (cf. Berkovitz [1974], Theorem 5.1 of Chapter III). Instead of the set $\hat{\mathbf{f}}(t, \mathbf{x}, \Omega)$ we consider the "super-set"

$$\mathcal{Q}^+(t, \mathbf{x}) = \{(y^0, \mathbf{y}^T) | \text{ for some } \mathbf{v} \in \Omega, \quad \mathbf{y} = \mathbf{f}(t, \mathbf{x}, \mathbf{v}), \quad y^0 \geq f^0(t, \mathbf{x}, \mathbf{v})\}.$$

EXAMPLE 5. Consider the scalar problem $\dot{x} = x + u$, $C[u(\cdot)] = \int_0^{t_1} [u(s)]^2 \, ds$. Then (Figure 6) for any (t, x):

$$\mathcal{Q}^+(t, \mathbf{x}) = \{(y^0, y) | y = x + v, \quad y^0 \geq v^2 \text{ for some } v \in \Omega\}.$$

Notice that $\hat{\mathbf{f}}(t, x, \Omega)$ is *not* convex while $\mathcal{Q}^+(t, x)$ is convex.

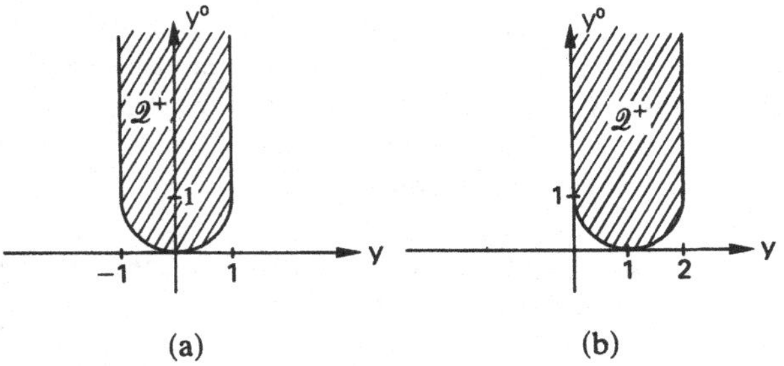

Figure 6 $\mathcal{Q}^+(t, x)$ with (a) $x = 0$, (b) $x = 1$

If $\hat{\mathbf{f}}(t, \mathbf{x}, \Omega)$ is convex, then so is $\mathcal{Q}^+(t, \mathbf{x})$. As the above example shows, however, the converse is not true. Therefore, the following theorem is more general than Theorem 2.

Theorem 3. *Consider the problem* (1) (2) *on a fixed compact interval* $[0, T]$. *Assume that* $\Delta(T) \neq \varnothing$ *and that all successful responses* $\mathbf{x}[\cdot]$ *satisfy an a priori bound* (3). *Assume that* $\mathbf{f}$ *is continuous,* f^0 *is lower semicontinuous, and that* $\mathcal{Q}^+(t, \mathbf{x})$ *is convex for* $(t, \mathbf{x}) \in [0, T] \times \{|\mathbf{x}| \leq \alpha\}$. *Then there exists an optimal control.*

This theorem can be substantially generalized, but a discussion of these generalizations would be a major digression. We refer the reader to Berkowitz [1974].

As an example of a practical class of problems of considerable interest to which Theorem 3 applies, we have the *linear plant with quadratic cost criterion*:

$$\dot{\mathbf{x}} = A(t)\mathbf{x} + B(t)\mathbf{u} + \mathbf{b}(t),$$

$$C[\mathbf{u}(\cdot)] = \lambda_1 \int_0^{t_1} |\mathbf{x}[t] - \mathbf{y}(t)|^2\, dt + \lambda_2 \int_0^{t_1} |\mathbf{u}(t)|^2\, dt,$$

where $\mathbf{y}(t)$ is given. The cost measures the mean square deviation of $\mathbf{x}[t]$ from a desired trajectory $\mathbf{y}(t)$ as well as the "fuel consumption" (assumed proportional to $|\mathbf{u}(t)|^2$). Theorem 2 does not apply to this problem – the situation is essentially that of Example 4(b), in which f^0 is quadratic, while $\mathbf{f}$ is linear in the control. However, Theorem 3 does apply to this problem.

5. Existence for Systems Linear in the State

In the following theorem, notice that there is *no* convexity assumption.

Theorem 4. *Consider the problem*

$$(5) \qquad \dot{\mathbf{x}} = A(t)\mathbf{x} + \mathbf{b}(t, \mathbf{u}), \qquad C[\mathbf{u}(\cdot)] = \int_0^{t_1} \{\mathbf{a}^T(t)\mathbf{x}[t] + \phi(t, \mathbf{u}(t))\}\, dt,$$

with A, $\mathbf{b}$, $\mathbf{a}$ and ϕ continuous $n \times n$, $n \times 1$, $n \times 1$ matrices and real-valued functions, respectively. As usual $\mathbf{x}_0$ is assumed fixed and $\mathcal{T}(t) \equiv \mathbf{0}$. If there exists a successful control ($\Delta \neq \varnothing$) and if successful responses satisfy an a priori bound (3), *then there exists an optimal control.*

Remark. We will only prove this theorem for the autonomous case:

$$\dot{\mathbf{x}} = A\mathbf{x} + \mathbf{b}(\mathbf{u}), \qquad C[\mathbf{u}(\cdot)] = \int_0^{t_1} \{\mathbf{a}^T\mathbf{x}[t] + \phi[\mathbf{u}(t)]\}\, dt,$$

with A, $\mathbf{b}$ and $\mathbf{a}$ constant. The proof for the non-autonomous case is essentially identical, except that it requires:

1. The Bang–Bang Principle for non-autonomous problems as stated just before Example 1 in Section 2 of Chapter II.
2. Certain of the extensions of Theorem 2 noted in the remarks following the proof of that theorem.

In our proof of the autonomous version of Theorem 4, we will use the Bang–Bang Principle in the following form (see the Introduction to Chapter III):

If $\mathcal{S} \subset R^m$ is any set, in R^n let $K_{\mathcal{S}}(t, \mathbf{x}_0)$ denote the reachable set for the system

$$(LA) \qquad \dot{\mathbf{x}} = A\mathbf{x} + B\mathbf{v}, \qquad \mathbf{x}(0) = \mathbf{x}_0, \qquad \mathbf{v}(t) \in \mathcal{S}, \quad \mathbf{v}(\cdot) \text{ measurable.}$$

Then $K_{\mathcal{S}}(t, \mathbf{x}_0) = K_{\text{co}(\mathcal{S})}(t, \mathbf{x}_0)$ *where* co $(\mathcal{S})$ *denotes the convex hull of* $\mathcal{S}$.

Proof of Theorem 4. We will study the extended response $\hat{\mathbf{y}}[t]$ for a slightly different system (we replace $\phi(u)$, $\mathbf{b}(u)$ by linear terms):

$$(\hat{5}) \qquad \hat{\mathbf{y}} = \begin{bmatrix} y^0 \\ \mathbf{y} \end{bmatrix}, \qquad \dot{y}^0 = \mathbf{a}^T\mathbf{y} + v^0(t), \qquad \dot{\mathbf{y}} = A\mathbf{y} + \mathbf{v}(t), \qquad \hat{\mathbf{v}} = \begin{bmatrix} v^0 \\ \mathbf{v} \end{bmatrix}$$

where $\hat{\mathbf{v}}(t)$ is constrained to belong to the convex hull of

$$\hat{\Omega} = \phi(\Omega) \times \mathbf{b}(\Omega) \quad \text{in } R^{n+1}.$$

The target is the set $\mathcal{T}(t) = \{[{\xi \atop \mathbf{0}}] \| \xi \in R\}$ in R^{n+1}, while $\hat{\mathbf{x}}_0 = [{0 \atop \mathbf{x}_0}]$.

We will show that $(\hat{5})$ satisfies the hypotheses of Theorem 2, with cost $\hat{C}[\hat{\mathbf{v}}(\cdot)] = y^0(t_1)$. Since there is by assumption a successful control $\mathbf{u}(t)$ for the original problem (5), there is a successful control $\hat{\mathbf{v}}^T(t) = (\phi(\mathbf{u}), \mathbf{b}^T(\mathbf{u}))$ for $(\hat{5})$. All responses $\hat{\mathbf{y}}[t]$ starting at $\hat{\mathbf{x}}_0$ satisfy an *a priori* bound, since the system $(\hat{5})$ is linear. The extended velocity set for $(\hat{5})$

$$\left\{\begin{bmatrix} \mathbf{a}^T\mathbf{y} + v^0 \\ A\mathbf{y} + \mathbf{v} \end{bmatrix} \middle| \hat{\mathbf{v}} \in \hat{\Omega}\right\},$$

is convex, since $(\hat{5})$ is linear in $\hat{\mathbf{v}}$ and $\hat{\mathbf{v}}(t)$ is restricted to a convex set.

Theorem 1 (with the easily verified Remark 8 following the proof) therefore may be applied to the problem $(\hat{5})$ with the given cost function $\hat{C}[\hat{\mathbf{v}}(\cdot)]$. We conclude that this problem has an optimal control, $\hat{\mathbf{v}}_*(t) \in$ co $(\hat{\Omega})$ with optimal response $\hat{\mathbf{x}}_*[t]$. By the Bang–Bang Principle there exists a control $\hat{\mathbf{v}}_*(t) \in \hat{\Omega}(t)$ with the same response, hence $\hat{\mathbf{v}}_*(t)$ is also optimal. Now $\hat{\mathbf{v}}_*(t) \in \phi(\Omega) \times \mathbf{b}(\Omega)$ for $0 \le t \le t_1$, so Fillipov's Lemma implies that there is a $\mathbf{u}_*(\cdot) \in \mathcal{U}_m$ such that $v_*^0(t) = \phi[\mathbf{u}_*(t)]$, $\mathbf{v}(t) = \mathbf{b}[\mathbf{u}_*(t)]$. The control $\mathbf{u}_*(\cdot)$ is easily seen to be optimal for (5). □

The extensions described in the Remarks following the proof of Theorem 1 are also valid for Theorem 4.

6. Applications

The Rocket Car

The most general optimal control problem formulated by us for the rocket car was:

$$\mathbf{x} = \begin{bmatrix} p \\ q \end{bmatrix}, \qquad \dot{\mathbf{x}} = \begin{bmatrix} 0 & 1 \\ 0 & 0 \end{bmatrix}\mathbf{x} + \begin{bmatrix} 0 \\ 1 \end{bmatrix} u(t),$$

$$C[u(\cdot)] = \int_0^{t_1} \{\lambda_1 + \lambda_2[q(t)]^2 + \lambda_3|u(t)|\}\, dt,$$

with $\lambda_i \geq 0$, $\sum_1^3 \lambda_i = 1$. We already know that any initial state in R^2 can be steered to the target $\mathbf{0}$, so for any fixed initial state, $\Delta(T)$ will be nonempty for T sufficiently large. Because the dynamical equations are linear in the state, all responses from a fixed initial state satisfy an *a priori* bound. Theorem 1 implies that there is an optimal control in each of the classes $\mathscr{U}_\lambda[0, T]$, $\mathscr{U}_r[0, T]$ for T sufficiently large. If $\lambda_1 \neq 0$, then

$$f^0(t, \mathbf{x}, u) \equiv \lambda_1 + \lambda_2[q(t)]^2 + \lambda_3|u| \geq \lambda_1,$$

and we can take $\eta(t) \equiv \lambda_1$ in Remark 4 to remove the restriction to $[0, T]$: there is an optimal control in each $\mathscr{U}_\lambda$, $\mathscr{U}_r$.

If $\lambda_2 = 0$, then Theorem 4 applies directly to yield the conclusion that there is an optimal control in $\mathscr{U}_m[0, T]$ for T large, and if $\lambda_1 > 0$ then we can again drop the restriction to $[0, T]$.

To see whether or not Theorems 2 and 3 apply to the rocket car, we must examine the sets (note that $\mathbf{f}(t, \mathbf{x}, u) = [\begin{smallmatrix} q \\ u \end{smallmatrix}]$):

$$\hat{\mathbf{f}}(t, \mathbf{x}, \Omega) = \{(\lambda_1 + \lambda_2[q(t)]^2 + \lambda_3|v|, q(t), v)^T \mid -1 \leq v \leq 1\}$$

and

$$\mathscr{Q}^+(t, x) = \{(v, q(t), u)^T \mid v \geq \lambda_1 + \lambda_2[q(t)]^2 + \lambda_3|u|, -1 \leq u \leq 1\}$$

in R^3. We can write

$$\hat{\mathbf{f}}(t, \mathbf{x}, \Omega) = \left\{\mathbf{c}(t, \mathbf{x}) + \begin{bmatrix} \lambda_3|u| \\ 0 \\ u \end{bmatrix} \middle| -1 \leq u \leq 1\right\}, \qquad \mathbf{c}(t, \mathbf{x}) = \begin{bmatrix} \lambda_1 + \lambda_2[q(t)]^2 \\ q(t) \\ 0 \end{bmatrix}.$$

Therefore, in (f^0, f^1, f^2)-space, $\mathbf{f}(t, \mathbf{x}, \Omega)$ is the translate by $\mathbf{c}(t, \mathbf{x})$ of the set $\{(\lambda_3|u|, 0, u) \mid -1 \leq u \leq 1\}$ – this set lies in the (f^0, f^2) plane (Figure 7(a)). $\mathscr{Q}^+(t, \mathbf{x})$ is just the translate by $\mathbf{c}(t, \mathbf{x})$ of the set $\{(v, 0, u) \mid v \geq \lambda_3|u|, -1 \leq u \leq 1\}$ (Figure 7(b)).

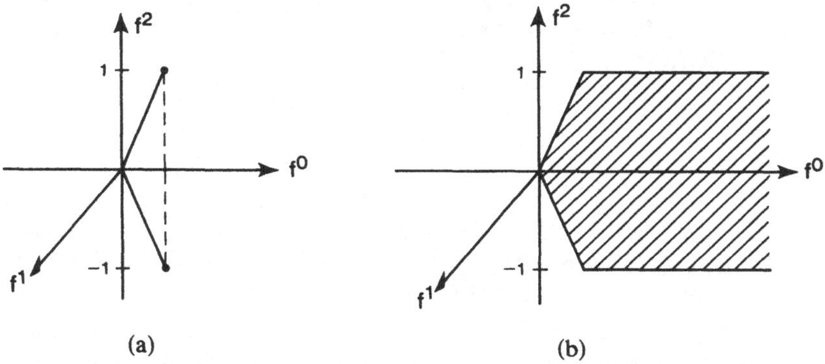

Figure 7 (a) $\hat{\mathbf{f}}(t, \mathbf{x}, \Omega)$ Is the Broken Line Segment, with Slopes $\pm 1/\lambda_3$. (b) $\mathscr{Q}^+(t, \mathbf{x})$ Is the 2-Dimensional Shaded Strip

From the sketches, we see that $\hat{\mathbf{f}}(t, x, \Omega)$ is only convex when $\lambda_3 = 0$, while $\mathcal{Q}^+(t, \mathbf{x})$ is always convex. So Theorem 2 applies only when we ignore fuel consumption ($\lambda_3 = 0$), while Theorem 3 covers all cases.

The Linearized Pendulum and the Physical Pendulum

In Chapter III (Example 2) we treated the time-optimal problem for a linearized pendulum, $\ddot{p} + kp = u$:

$$\mathbf{x} = \begin{bmatrix} p \\ q \end{bmatrix}, \qquad \dot{\mathbf{x}} = \begin{bmatrix} 0 & 1 \\ -k & 0 \end{bmatrix}\mathbf{x} + \begin{bmatrix} 0 \\ 1 \end{bmatrix} u, \qquad (k > 0), \qquad C[u(\cdot)] = \int_0^{t_1} 1\, dt = t_1.$$

Any of our theorems can be applied to assert the existence of an optimal control. We are not restricted to a compact interval, since $f^0(t, \mathbf{x}, u) \geq 1$ and $\int^\infty 1 = +\infty$.

The physical pendulum is actually governed by the nonlinear equation $\ddot{p} + \beta\dot{p} + k \sin p = u(t)$, with $\beta > 0$ representing the damping, so the time-optimal problem is:

$$\mathbf{x} = \begin{bmatrix} p \\ q \end{bmatrix}, \qquad \dot{\mathbf{x}} = \begin{bmatrix} q \\ -\beta q - k \sin p \end{bmatrix} + \begin{bmatrix} 0 \\ 1 \end{bmatrix} u,$$

with $C[u(\cdot)]$ as above. Responses are bounded on any compact interval, since $\mathbf{f}$ is bounded by a linear function:

$$|\mathbf{f}(t, \mathbf{x}, u)| \leq [|q| + |\beta q + k \sin p|] + |u| \leq k_* |\mathbf{x}| + 1, \qquad k_* = \max(\beta, k, 1).$$

In this case, $\hat{\mathbf{f}}(t, \mathbf{x}, \Omega)$ is convex, since it is just a line segment:

$$\begin{aligned} \hat{\mathbf{f}}(t, \mathbf{x}, \Omega) &= \{(1, q, -\beta q - k \sin p + u)^T | -1 \leq u \leq 1\} \\ &= \mathbf{c}(t, \mathbf{x}) + \{(0, 0, u)^T | -1 \leq u \leq 1\} \end{aligned}$$

where $\mathbf{c}(t, \mathbf{x}) = (1, q, -\beta q - k \sin p)^T$. Therefore, Theorem 2 applies *if* we can show that $\Delta(T) \neq 0$ for the given initial state and some $T > 0$. This is indeed possible for any initial state, but a proof would be too lengthy here (see Lee and Markus [1967], p. 446).

Exercises

1. Show that if $|\mathbf{x}^T\mathbf{f}(t, \mathbf{x}, \mathbf{v})| \leq \alpha\|\mathbf{x}\|^2 + \beta$ for all $0 \leq t \leq T$, $\mathbf{x} \in R^n$, $\mathbf{v} \in \Omega$, then any solution of $\dot{\mathbf{x}} = f(t, \mathbf{x}, \mathbf{u}(t))$, $\mathbf{u}(\cdot) \in \mathcal{U}_m[0, T]$, with $\mathbf{x}(0) = \mathbf{x}_0$ fixed, satisfies an *a priori* bound (3). (*Hint*: What is $d(\mathbf{x}^T\mathbf{x})/dt$?)

2. In Example 2 we considered a particular system linear in the state, and showed that there was no optimal control. Explain why each of Theorems 2–4 does not apply. Explain why our presentation of this counterexample does not go through when we are restricted to the control classes $\mathcal{U}_\lambda$, $\mathcal{U}_r$ (Theorem 1).

3. Consider the scalar problem $\dot{x} = u$, $x(0) = 0$, with target $\mathscr{T}(t) \equiv 1$, the control class $\{u(\cdot)$ measurable, $0 \le u(t) < \infty\}$, and cost $C[u(\cdot)] = \int_0^{t_1} [x[s]]^2\, ds$. Show that there is no optimal control. (*Hint*: Show that the optimal cost is zero, by constructing $\{u_k(\cdot)\}$ for which $C[u_k(\cdot)] \to 0$). Explain why our theorems (including the extensions noted in the remarks) do not apply.

4. Consider the scalar problem $\dot{x} = u$, $x(0) = 0$, $\mathscr{T}(t) \equiv 1$, with the control class $\{u(\cdot)$ measurable, $0 \le u(t) < 1\}$, and cost $C[u(\cdot)] = t_1 = \int_0^{t_1} dt$. Show that there is no optimal control, and explain why our theorems don't apply. (*Hint*: Note that $x(t_1) = 1 = \int_0^{t_1} u(t)\, dt$. Show that the optimal time t_1 is 1.)

5. Consider the scalar problem $\dot{x} = u$, with $x(0) = 0$, the usual control class $\mathscr{U}_m$, the target set

$$\mathscr{T}(t) = \{x \mid 0 < x \le 1\},$$

and cost $C[u(\cdot)] = \int_0^{t_1} (x[t])^2\, dt$. Show that no optimal control exists, and explain why our theorems do not apply. (*Hint*: Let $u_k(t) \equiv 1/k$ on $0 \le t \le 1$.)

6. Consider $\dot{x} = u$, $x(0) = 0$, $\mathscr{T}(t) \equiv 1$ with control class $\{u(\cdot)$ measurable, $0 \le u(t) \le 1\}$, and cost $C[u(\cdot)] = \int_0^{t_1} [u(t) + 1] e^{-t}\, dt$. Show that there is no optimal control and explain why our theorems do not apply. (*Hint*: Let $u_k(t) = 1/k$ on $[0, k]$).

7. Consider the system ($n = 3$, $m = 2$)

$$\dot{\mathbf{x}} = \begin{bmatrix} 0 \\ 0 \\ 1 \end{bmatrix} + \begin{bmatrix} 1 & 0 \\ 0 & 1 \\ 0 & 0 \end{bmatrix} \mathbf{u}, \qquad \mathbf{x} = \begin{bmatrix} p \\ q \\ r \end{bmatrix}, \qquad \mathbf{u} = \begin{bmatrix} u^1 \\ u^2 \end{bmatrix}$$

with

$$\mathbf{x}(0) = \begin{bmatrix} 0 \\ 0 \\ -1 \end{bmatrix}, \qquad \mathscr{T}(t) \equiv \begin{bmatrix} 0 \\ 0 \\ 0 \end{bmatrix}, \qquad C[\mathbf{u}(\cdot)] = \int_0^{t_1} \{\|\mathbf{x}\|^2 + [1 - \|\mathbf{u}\|^2]^2\}\, dt.$$

We use our usual control class $\mathscr{U}_m$, and remind the reader that $\|\mathbf{x}\|^2 = p^2 + q^2 + r^2$. Show that $\{\mathbf{u}_k(t)\} = \{(\sin 2\pi kt, \cos 2\pi kt)^T\}$ is a minimizing sequence. Show that there is no optimal control, and explain why our theorems do not apply.

8. Consider the problem ($n = 2$, $m = 1$)

$$\mathbf{x} = \begin{bmatrix} p \\ q \end{bmatrix}, \qquad \dot{p} = -[q]^2 + u, \qquad \dot{q} = u$$

with our usual control class $\mathscr{U}_m$, $\mathbf{x}_0 = (0, 0)^T$,

$$\mathscr{T}(t) = \{(1, q) \mid -1 \le q \le 1\},$$

and $C[u(\cdot)] = \int_0^{t_1} ds = t_1$. Sketch $\mathscr{Q}^+(t, \mathbf{x})$ in R^3. Show that, although $\mathscr{Q}^+$ is convex, there is no successful control.

9. (Continuing problem (8) from Chapter I). Consider the scalar problem

$$\frac{dx}{dt} = rx\left(1 - \frac{x}{k}\right) - u(t), \qquad x(0) = x_0$$

on $[0, T]$, with r, k constant and $u(\cdot)$ measurable, $0 \leq u(t) \leq 1$, and cost $C[u(\cdot)] = \int_0^T e^{-\delta t}[p - c(\dot{x}[t])]\, dt$ where $\delta > 0$, $p > 0$ are given constants and $c(v) \geq 0$ is continuous. Discuss conditions on $c(\cdot)$ under which the theorems of this chapter can be applied. Pay special attention to the control classes $\mathscr{U}_\lambda$, $\mathscr{U}_r$.

10. Prove Remark 4, which follows the proof of Theorem 1. (*Hint*: If $\Delta \neq \varnothing$, then there is a successful control $\mathbf{u}(\cdot)$ with cost α. Show that for t_1 sufficiently large, a successful control which hits the target at time t_1 will have a cost exceeding α.)

11. (See the discussion following the statement of Theorem 3.) Show that for the linear plant with quadratic cost, Theorem 2 does not apply, while Theorem 3 asserts the existence of an optimal control.

Notes

Theorem 1 for $\mathscr{U}_\lambda$ is due to Lee and Markus; for $\mathscr{U}_r$ it is due to Strauss [1968]. Theorem 2 is due to Fillipov [1959] and Roxin [1962]. Theorem 3 is due to Cesari – for a full discussion of this type of result, see Cesari [1971] and Berkovitz [1974].

Remark 4 following the proof of Theorem 1, to the effect that we can work on $[0, \infty)$ or $(-\infty, \infty)$ if we assume a high cost for controls which go on for too long, was first observed by Russell [1964]. He also used the interesting concept of a *penalty function*, associating high cost with leaving a prescribed $(t, \mathbf{x}, \mathbf{u})$-set.

Theorem 4 was first stated by Neustadt, and extended by Olech [1966] and Jacobs [1968]. For a full discussion of this theorem, see Section 6 of Chapter 4 of Berkovitz [1974].

For a survey of the state of existence theory as of 1978, see Berkovitz [1978].

Chapter V

Necessary Conditions for Optimal Controls – The Pontryagin Maximum Principle

1. Introduction

In Chapter IV we described conditions which guarantee the existence of at least one optimal control – we call these *sufficient conditions*. However, these sufficient conditions are not very helpful in actually finding an optimal control. In this chapter we will describe one rather complicated set of conditions which any optimal control must necessarily satisfy. This set of *necessary conditions* is collectively known as the Pontryagin Maximum Principle (PMP). For many important problems, the conditions of the PMP will only be satisfied by a small subset of our control class (perhaps only by a single control). In this case there is a reasonable chance of our finding an optimal control *if one exists*. We emphasize that the PMP is a *necessary* set of conditions – there may be no optimal control, yet the PMP may delineate a nonempty class of candidates. To be sure that an optimal control actually exists, we must appeal to sufficiency theorems from Chapter IV.

In Section 2 we describe the PMP for autonomous problems. Because a complete proof is very technical, we have placed ours in the Appendix to this chapter. Section 3 is entirely devoted to examples illustrating the application of the PMP. In Section 4 we give a heuristic proof of the PMP which shows how the concept of costate arises naturally, and in Section 5 we give two important extensions, one to non-autonomous problems, and the other to problems involving sets as initial and target data:

$$\begin{Bmatrix} \mathbf{x}[t_0] = \mathbf{x}_0 \\ \mathbf{x}[t_1] = \mathbf{x}_1 \end{Bmatrix} \quad \text{is replaced by} \quad \begin{Bmatrix} \mathbf{x}[t_0] \in S_0 \\ \mathbf{x}[t_1] \in S_1 \end{Bmatrix},$$

where S_0 and S_1 are specified sets.

Throughout this chapter we use measurable controls, the Lebesgue integral, and the phrase "almost everywhere" (a.e.) occurs frequently. The reader may use instead piecewise smooth or piecewise continuous controls with the Riemann integral. This involves replacing the phrase "a.e." by the phase "at all but a finite set of points." The proofs are easily modified – in fact, often simplified.

2. The Pontryagin Maximum Principle for Autonomous Systems

We consider the autonomous problem:

(1) $$\dot{\mathbf{x}} = \mathbf{f}(\mathbf{x}, \mathbf{u}), \qquad \mathbf{x}[t] \in R^n, \qquad \mathbf{u}(t) \in R^m$$

with fixed initial instant and state $\mathbf{x}[t_0] = \mathbf{x}_0$, fixed target state $\mathcal{T}(t) \equiv \mathbf{x}_1$, and cost

$$C[\mathbf{u}(\cdot)] = \int_{t_0}^{t_1} f^0(\mathbf{x}[t], \mathbf{u}(t))\, dt.$$

Here t_1 is the unspecified time of arrival at the target state, $\mathbf{x}[t_1] = \mathbf{x}_1$, and f^0 is assumed given. *We use the control set*

$$\mathbf{u}(\cdot) \in \mathcal{V}_m \equiv \{\mathbf{u}(\cdot) | \mathbf{u}(\cdot) \text{ measurable, } \mathbf{u}(t) \in \Psi \text{ for } t_0 \leq t \leq t_*\}$$

where Ψ is a given bounded set in R^m and t_* depends on $\mathbf{u}(\cdot)$. We will always assume that f and f^0 are continuous in $(\mathbf{x}, \mathbf{u})$ and continuously differentiable in $\mathbf{x}$. Notice that the set in which $\mathbf{u}(\cdot)$ takes its values need not be convex (nor open, nor need $\mathbf{0}$ belong to the set). For example,

$$\Psi = \left\{ \begin{bmatrix} u^1 \\ u^2 \end{bmatrix} \middle| |u^1| + |u^2| = 1 \right\}$$

is allowed.

Before we can even state the PMP we need some notation and analysis. For a given control $\mathbf{u}(\cdot)$ and (any) associated response $\mathbf{x}[\cdot]$, we define the "dynamic cost" variable

$$x^0[t] = \int_{t_0}^{t} f^0(\mathbf{x}[s], \mathbf{u}(s))\, ds.$$

If $\mathbf{u}(\cdot)$ is successful, then $\mathbf{x}[t_1] = \mathbf{x}_1$ for some $t_1 \geq t_0$, and the associated cost is $x^0[t_1]$. If $\mathbf{u}(\cdot)$ is optimal, then $x^0[t_1]$ is as small as possible.

We increase the dimension of our problem by defining the $(n+1)$-vector $\hat{\mathbf{x}}[t] = (x^0[t], \mathbf{x}^T[t])^T$. If we set $\hat{\mathbf{f}}(t, \hat{\mathbf{x}}) = (f^0, \mathbf{f}^T)^T$, then our original problem can be restated as follows:

Find an admissible control $\mathbf{u}(\cdot)$ *such that the* $(n+1)$*-dimensional solution of*

$$(\hat{1}) \qquad \dot{\hat{\mathbf{x}}}[t] = \hat{\mathbf{f}}(\hat{\mathbf{x}}[t], \mathbf{u}(t)) \text{ a.e.}, \qquad \hat{\mathbf{x}}[t_0] = \begin{bmatrix} 0 \\ \mathbf{x}_0 \end{bmatrix},$$

terminates at

$$\begin{bmatrix} x^0[t_1] \\ \mathbf{x}_1 \end{bmatrix} \quad (\mathbf{x}_1 \textit{ the given target state})$$

with $x^0[t_1]$ *as small as possible. Again we emphasize that* t_1 *is not specified.*

In other words, we want the extended state vector $\hat{\mathbf{x}}[t]$ to hit the line

$$\hat{\mathcal{T}}(t) = \{(\xi, \mathbf{x}_1) | \xi \text{ real, } \mathbf{x}_1 \text{ the given target}\}$$

as far down the x^0-axis as possible. Figure 1 represents a two-dimensional problem ($n = 2$) extended to three dimensions. We can only require $(\hat{1})$ to be satisfied a.e. because $\hat{\mathbf{f}}(\hat{\mathbf{x}}, \mathbf{u}(t))$ is only measurable in t (cf. the Mathematical Appendix).

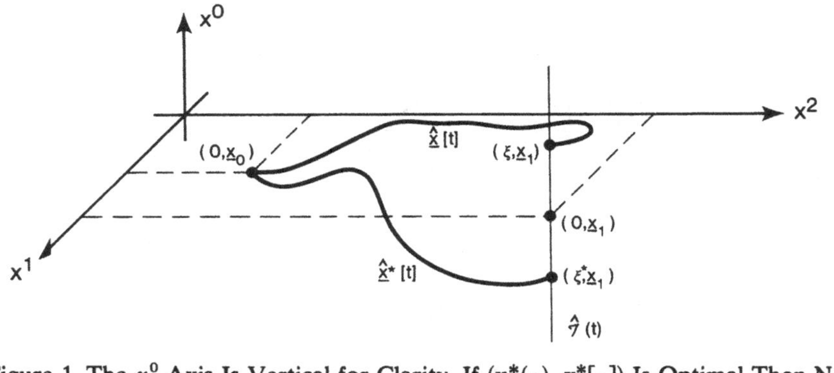

Figure 1 The x^0 Axis Is Vertical for Clarity. If $(\mathbf{u}^*(\cdot), \mathbf{x}^*[\cdot])$ Is Optimal Then No Extended Response Can Hit the Line $\hat{\mathcal{T}}(t)$ Below ξ^*

Having extended the dimension of our original problem, we now turn to a rather complicated animal, the *adjoint to the linearization of* $(\hat{1})$:

For a given admissible control $\mathbf{u}(\cdot)$ *and associated (extended) response* $\hat{\mathbf{x}}[\cdot]$, *we consider the* $(n+1)$*-dimensional linear system*

$$(\widehat{\text{Adj.}}) \qquad \dot{\hat{\mathbf{w}}}(t) = -\hat{\mathbf{f}}_{\hat{\mathbf{x}}}(\mathbf{x}[t], \mathbf{u}(t))^T \hat{\mathbf{w}}(t), \quad \text{a.e.}$$

The solutions of this system are called *extended costates.* Here $\hat{\mathbf{f}}_{\hat{\mathbf{x}}}(\mathbf{x}[t], \mathbf{u}(t)) = \partial \hat{f}^i / \partial \hat{x}^j$; $i, j = 0, 1, \ldots, n$ is the Jacobian matrix of $\hat{\mathbf{f}}$ with respect to $\hat{\mathbf{x}}$. Because none of the $\hat{f}^i$'s involves x^0, this Jacobian has the form

$$\hat{\mathbf{f}}_{\hat{\mathbf{x}}}(\mathbf{x}[t], \mathbf{u}(t)) = \left[\frac{\partial \hat{f}^i}{\partial \hat{x}^j}\right] = \begin{bmatrix} 0 & \partial f^0/\partial x^1 & \cdots & \partial f^0/\partial x^n \\ 0 & \partial f^1/\partial x^1 & \cdots & \partial f^1/\partial x^n \\ \vdots & & & \\ 0 & \partial f^n/\partial x^1 & \cdots & \partial f^n/\partial x^n \end{bmatrix}.$$

Before we continue, we discuss for the reader's benefit a geometric interpretation of $(\widehat{\text{Adj}}.)$ The details are outlined in Exercises 1 and 2 at the end of the chapter. For a given constant control $\mathbf{u}(\cdot)$, any solution $\hat{\mathbf{x}}[\cdot]$ of $\dot{\hat{\mathbf{x}}}[t] = \hat{\mathbf{f}}(\mathbf{x}[t], \mathbf{u}(t))$ is a curve in R^{n+1}. If $\hat{\mathbf{b}}_0$ is a tangent vector to $\hat{\mathbf{x}}[\cdot]$ at $\hat{\mathbf{x}}[t_0]$, then the solution $\hat{\mathbf{b}}(t)$ of the *linearized* equation:

$(\widehat{\text{Lin}})$ $\quad \dot{\hat{\mathbf{b}}}(t) = \hat{\mathbf{f}}_{\hat{\mathbf{x}}}(\mathbf{x}[t], \mathbf{u}(t))\hat{\mathbf{b}}(t), \quad \hat{\mathbf{b}}(t_0) = \hat{\mathbf{b}}_0 \quad u(t)$ constant,

will be tangent to this curve at $\hat{\mathbf{x}}[t]$ for all t. Thus the linearized equation describes the evolution of tangent vectors along the solution curves of the resulting autonomous equation (Figure 2).

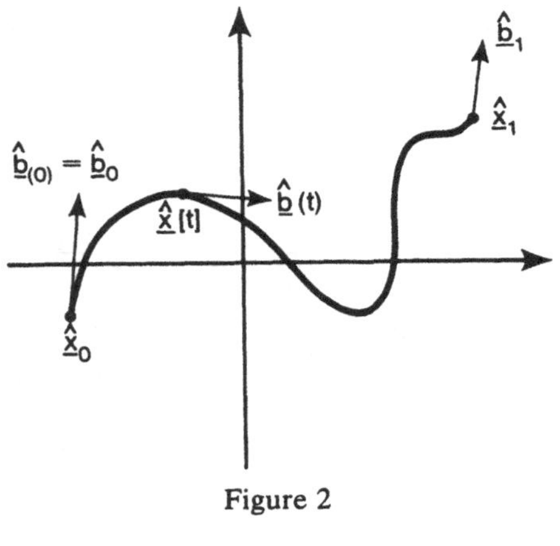

Figure 2

EXAMPLE 1. For $u(t) \equiv +1$, the three-dimensional problem

$$\dot{x}^0 = x^1, \qquad \dot{x}^1 = -ux^0, \qquad \dot{x}^2 = u - 1, \qquad \hat{\mathbf{x}}[0] = \begin{bmatrix} 0 \\ 1 \\ 0 \end{bmatrix},$$

has response

$$x^0[t] = \sin t, \qquad x^1[t] = \cos t, \qquad x^2(t) \equiv 0.$$

Then

$$\hat{\mathbf{f}}(\mathbf{x}, u) = \begin{bmatrix} x^1 \\ -ux^0 \\ u-1 \end{bmatrix}, \qquad \hat{\mathbf{f}}_{\hat{\mathbf{x}}}(\mathbf{x}[t], u(t)) = \begin{bmatrix} 0 & 1 & 0 \\ -u & 0 & 0 \\ 0 & 0 & 0 \end{bmatrix} = \begin{bmatrix} 0 & 1 & 0 \\ -1 & 0 & 0 \\ 0 & 0 & 0 \end{bmatrix}.$$

Therefore the linearized equation is:

$$\dot{\hat{\mathbf{b}}}(t) = \begin{bmatrix} 0 & 1 & 0 \\ -1 & 0 & 0 \\ 0 & 0 & 0 \end{bmatrix} \hat{\mathbf{b}}(t).$$

Clearly, the vector $\hat{\mathbf{b}}_0 = (1, 0, 0)^T$ is tangent to $\hat{\mathbf{x}}[\cdot]$ at $\hat{\mathbf{x}}[0]$ (Figure 3), and the solution of the linearized equation with $\hat{\mathbf{b}}(0) = \mathbf{b}_0$ is $\hat{\mathbf{b}}(t) = (\cos t, -\sin t, 0)^T$. Since $\hat{\mathbf{x}}[t]$ is the radius vector of a circle, and $\hat{\mathbf{b}}(t)$ is perpendicular to $\hat{\mathbf{x}}[t]$ for all t, it follows that $\mathbf{b}(t)$ is always tangent to this circle.

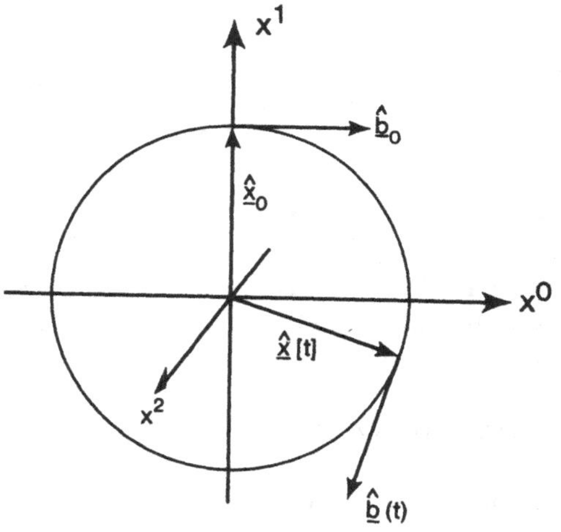

Figure 3

Now the significance of $(\widehat{\text{Adj.}})$ is that $\langle \hat{\mathbf{w}}(t), \hat{\mathbf{b}}(t) \rangle \equiv$ constant for any solutions of $(\widehat{\text{Adj.}})$ and $(\widehat{\text{Lin.}})$, respectively, (cf. Exercise 2). Thus, if $\hat{\mathbf{b}}(t)$ is tangent to $\hat{\mathbf{x}}[\cdot]$ at $\hat{\mathbf{x}}[t]$ for all t, and if $\hat{\mathbf{w}}(t_0)$ is perpendicular to $\mathbf{b}(t_0)$, then $\hat{\mathbf{w}}(t)$ will be perpendicular to $\hat{\mathbf{x}}[\cdot]$ at $\hat{\mathbf{x}}[t]$ for all t.

Thus $(\widehat{Adj.})$ describes the evolution of vectors lying in the n-dimensional hyperplane $P(t)$ attached to the extended response curve $\hat{\mathbf{x}}[\cdot]$ (Figure 4).

When $\mathbf{u}(t)$ is not identically constant, the situation is more complicated, and we discuss it in the Appendix to this chapter. In this case, $(\widehat{\text{Lin}})$ does not in general describe the evolution of a tangent vector to $\hat{\mathbf{x}}(\cdot)$. Instead, it describes a first approximation to the evolution of perturbations in the state—if you perturb $\hat{\mathbf{x}}[0]$ to $\hat{\mathbf{x}}_0 + \hat{\boldsymbol{\delta}}$, the response at a later time t is approximately changed from $\hat{\mathbf{x}}[t]$ to $\hat{\mathbf{x}}[t] + \hat{\mathbf{b}}(t)$, where $\hat{\mathbf{b}}(\cdot)$ solves $(\widehat{\text{Lin}})$ with $\hat{\mathbf{b}}(0) = \hat{\boldsymbol{\delta}}$. $(\widehat{\text{Adj}})$ will then describe the evolution of vectors in the hyperplane which moves along $\hat{\mathbf{x}}[t]$ with normal $\hat{\mathbf{b}}(t)$.

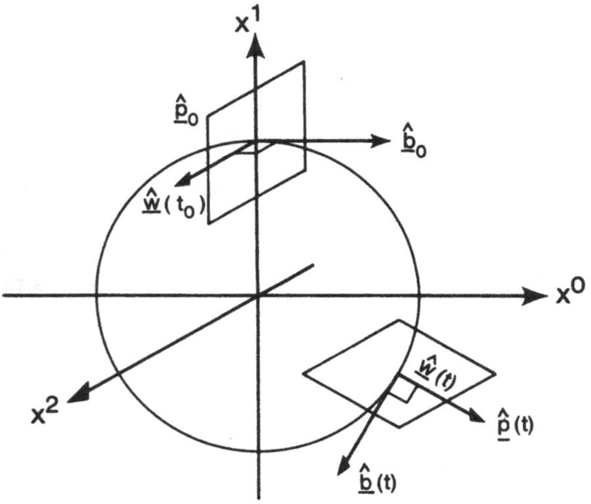

Figure 4

Before we state the PMP, we need one more concept. For a given control and (extended) response $(\hat{\mathbf{x}}[\cdot], \mathbf{u}(\cdot))$, we take any costate $\hat{\mathbf{w}}(\cdot)$ and form the real-valued function of time:

$$(2) \qquad H(\hat{\mathbf{w}}, \hat{\mathbf{x}}, \mathbf{u}) \equiv \langle \hat{\mathbf{w}}, \hat{\mathbf{f}} \rangle = \sum_{j=0}^{n} w^j(t) f^j(\mathbf{x}[t], \mathbf{u}(t)).$$

Then H is a *Hamiltonian* for $(\hat{1})$ and $(\widehat{\text{Adj}}.)$, i.e., we can write:

$$(\hat{1}) \qquad \dot{\hat{\mathbf{x}}} = \text{grad}_{\hat{\mathbf{w}}} H(\hat{\mathbf{w}}, \hat{\mathbf{x}}, \mathbf{u}) = \left(\frac{\partial H}{\partial w^0}, \frac{\partial H}{\partial w^1}, \ldots, \frac{\partial H}{\partial w^n}\right)^T, \quad \text{a.e.;}$$

$$(\widehat{\text{Adj}}.) \qquad \dot{\hat{\mathbf{w}}} = -\text{grad}_{\hat{\mathbf{x}}} H(\hat{\mathbf{w}}, \hat{\mathbf{x}}, \mathbf{u}) = -\left(\frac{\partial H}{\partial x^0}, \frac{\partial H}{\partial x^1}, \ldots, \frac{\partial H}{\partial x^n}\right)^T, \quad \text{a.e.}$$

The concept of Hamiltonian is from the science of mechanics (a discussion can be found in Pontryagin *et al.* [1964]). The function H does not depend on x^0 (since no f^j depends on x^0), so we write $H(\hat{\mathbf{w}}, \mathbf{x}, \mathbf{u})$.

We now look at $H = \sum_{j=0}^{n} w^j f^j(\mathbf{x}, \mathbf{v})$ *as a function of arbitrary vectors* $\hat{\mathbf{w}}$, $\mathbf{x}$, and $\mathbf{v} \in \Psi$. For fixed vectors $\mathbf{x} \in R^n$ and $\hat{\mathbf{w}} \in R^{n+1}$ we define

$$M(\hat{\mathbf{w}}, \mathbf{x}) = \sup_{\mathbf{v} \in \Psi} H(\hat{\mathbf{w}}, \mathbf{x}, \mathbf{v}).$$

In plain English, M is the largest value of H you can get for the given vectors $(\hat{\mathbf{w}}, \mathbf{x})$, using admissible values for $\mathbf{v}$. And now, finally,

Theorem 1 (The Pontryagin Maximum Principle). *Consider the extended control problem* $(\hat{1})$ *with measurable controls* $\mathbf{u}(\cdot)$ *taking values in a fixed*

bounded set $\Psi \subset R^m$. *Suppose* $(\mathbf{u}(\cdot), \hat{\mathbf{x}}[\cdot])$ *is an optimal control-response pair. Then there exists an absolutely continuous function* $\hat{\mathbf{w}}(\cdot)$ *solving* $(\widehat{Adj}.)$ a.e. *on* $[t_0, t_1]$, *with*

(i) $$H(\hat{\mathbf{w}}(t), \mathbf{x}[t], \mathbf{u}(t)) = M(\hat{\mathbf{w}}(t), \mathbf{x}[t]) \quad \text{a.e.},$$

(ii) $$M(\hat{\mathbf{w}}(t), \mathbf{x}[t]) \equiv 0 \quad \textit{on } [t_0, t_1],$$

(iii) $$w^0(t) \equiv w^0(t_0) \le 0 \quad \textit{and} \quad \hat{\mathbf{w}}(t) \ne \hat{\mathbf{0}} \quad \textit{on } [t_0, t_1].$$

The principle says that if $\mathbf{u}(\cdot)$ is optimal for $(\hat{1})$, then there is an associated response-adjoint pair, $(\hat{\mathbf{x}}[\cdot], \hat{\mathbf{w}}(\cdot))$, such that for each instant t, $H(\hat{\mathbf{w}}(t), \mathbf{x}[t], \mathbf{v}) \le 0$ for any $\mathbf{v} \in \Psi$, and equality is attained for $\mathbf{v} = \mathbf{u}(t)$ (it may be attained for other values of $\mathbf{v}$ as well). It is this "maximum" part of the theorem that will often single out a small class of candidates for the optimal control. We emphasize there may be a nonempty set of candidates, yet no optimal control for a given problem – the Pontryagin Maximum Principle *assumes* that an optimal control exists. This is why we need the sufficient conditions of Chapter IV.

Remark. Notice that the Pontryagin Maximum Principle applies to a *minimization* problem. If instead, one wishes to maximize a cost

$$J[\mathbf{u}(\cdot)] = \int_0^{t_1} g(\mathbf{x}[s], \mathbf{u}(s))\, ds,$$

then one would apply the PMP to the problem of minimizing $C[\mathbf{u}(\cdot)] = -J[\mathbf{u}(\cdot)]$, so $f^0 = -g$.

EXAMPLE 2. To help the reader become more familiar with the notation and concepts used in the PMP, we will work through some of the details of its application to the time-optimal problem for the Rocket Car:

$$\mathbf{x} = \begin{bmatrix} p \\ q \end{bmatrix}, \qquad \hat{\mathbf{x}} = \begin{bmatrix} x^0 \\ p \\ q \end{bmatrix}, \qquad C[u(\cdot)] = \int_0^{t_1} 1\, dt = t_1,$$

$$\dot{x}^0 = 1, \qquad \dot{p} = q, \qquad \dot{q} = u, \qquad \hat{\mathbf{f}}(\mathbf{x}, u) = \begin{bmatrix} 1 \\ q \\ u \end{bmatrix},$$

$$\hat{\mathbf{f}}_{\hat{\mathbf{x}}} = \begin{bmatrix} 0 & 0 & 0 \\ 0 & 0 & 1 \\ 0 & 0 & 0 \end{bmatrix}, \qquad \hat{\mathbf{w}} = \begin{bmatrix} w^0 \\ w^1 \\ w^2 \end{bmatrix},$$

$$\dot{w}^0 = \dot{w}^1 = 0, \qquad \dot{w}^2 = -w^1,$$

$$H(\hat{\mathbf{w}}, \mathbf{x}, v) = \langle \hat{\mathbf{f}}, \hat{\mathbf{w}} \rangle = w^0(0) + w^1(0) q + [w^2(0) - w^1(0) t] v.$$

The PMP asserts that if $u(\cdot)$ is time-optimal, then there are numbers $w^i(0)$, $i=0, 1, 2$, such that $H \leq 0$ for all $v \in \Psi$, and $H=0$ a.e. when $v=u(t)$. For example if $\Psi = \Omega \equiv [-1, 1]$, then clearly

$$M \equiv \max_{v\in\Psi} H = w^0(0) + w^1(0)q[t] + |w^2(0) - w^1(0)t|,$$

and this is clearly only attained for

$$u(t) = \operatorname{sgn}[w^2(0) - w^1(0)t].$$

Now the linear function $w^2(0) - w^1(0)t$ is either identically zero, or vanishes exactly once; thus *if a time-optimal control exists* it is almost everywhere either identically zero or bang–bang. In fact, it cannot be identically zero, since if it were the case that $w^1(0) = w^2(0) = 0$, then we would have $w^0(0) \neq 0$ ($\hat{\mathbf{w}}(t)$ can never vanish), which would imply $H \equiv w^0(0) \neq 0$ and so $M = \max_{v\in\Psi} H = w^0(0) \neq 0$ contradicting assertion (ii) of the PMP. Notice that all of the above conclusions were derived without knowing the explicit form of $\hat{\mathbf{w}}(t)$.

Before we further illustrate the use of the PMP, we need to reduce the nuisance caused by the "a.e." in part (i).

Lemma 1. *The equality* (i) $H(\hat{\mathbf{w}}(t), \mathbf{x}[t], \mathbf{u}(t)) = M(\hat{\mathbf{w}}(t), \mathbf{x}[t])$ (a.e.) *from the PMP is valid at every point t for which there is a sequence $t_k \to t$ with $H=M$ at each t_k and $\mathbf{u}(t_k) \to \mathbf{u}(t)$. In particular,* (i) *holds at every point of left or right continuity of* $\mathbf{u}(\cdot)$.

Proof. Choose $t_k \to t$ with $\mathbf{u}(t_k) \to \mathbf{u}(t)$ and $H=M$ at each t_k. For any $\mathbf{v} \in \Psi$

$$H(\hat{\mathbf{w}}(t_k), \mathbf{x}[t_k], \mathbf{u}(t_k)) \geq H(\hat{\mathbf{w}}(t_k), \mathbf{x}[t_k], \mathbf{v}).$$

Now $\hat{\mathbf{w}}(\cdot)$ and $\mathbf{x}[\cdot]$ are continuous, and H is a continuous function of $(\hat{\mathbf{w}}, \mathbf{x}, \mathbf{u})$, therefore we let $t_k \to t$ to obtain

$$H(\hat{\mathbf{w}}(t), \mathbf{x}[t], \mathbf{u}(t)) \geq H(\hat{\mathbf{w}}(t), \mathbf{x}[t], \mathbf{v}).$$

Since $\mathbf{v} \in \Psi$ was arbitrary, the result follows. □

Corollary. *Consider the optimization problem described in the PMP, with $\mathbf{u}(\cdot)$ restricted to the class $\mathscr{V}_{PS}$ of piecewise smooth functions with values in Ψ. If $(\mathbf{u}(\cdot), \hat{\mathbf{x}}[\cdot])$ is an optimal pair, then there exists a continuous costate $\hat{\mathbf{w}}(\cdot)$, never $\hat{\mathbf{0}}$, with piecewise continuous derivative, such that*

$$\dot{\hat{\mathbf{w}}} = -[\hat{\mathbf{f}}_{\hat{\mathbf{x}}}]^T \hat{\mathbf{w}}$$

holds except perhaps for finitely many points (and at those points, the one-sided limits of $\dot{\hat{\mathbf{w}}}$ and $-[\hat{\mathbf{f}}_{\hat{\mathbf{x}}}]^T\hat{\mathbf{w}}$ exist and are equal). In addition, (i), (ii), *and* (iii) *of the PMP hold* everywhere. *The same conclusion holds if we replace* $\mathscr{V}_{PS}$ *by* $\mathscr{V}_{PCN}$, *the piecewise continuous control functions with values in* Ψ.

Proof. The only assertion which is not obvious is the claim that $\dot{\hat{\mathbf{w}}} = -[\hat{\mathbf{f}}_{\hat{\mathbf{x}}}]^T \hat{\mathbf{w}}$ at all but a finite number of points. This stems directly from the method for solving an equation of the form $\dot{\mathbf{y}} = A(t)\mathbf{y}$ with $A(t)$ piecewise continuous. Specifying a value of $\mathbf{y}(t_0) = \mathbf{y}_0$ for some t_0, this differential equation can be solved on each interval $[a_n, b_n]$ over which $A(t)$ is continuous, using the value $\mathbf{y}(b_n)$ for the initial value on the next interval. This solution is clearly piecewise smooth, and $\dot{\mathbf{y}} = A(t)\mathbf{y}$ everywhere except at the discontinuities of A (where the one-sided limits are equal).

To show that $A(t) \equiv -[\hat{\mathbf{f}}_{\hat{\mathbf{x}}}(\mathbf{x}[t], \mathbf{u}(t))]^T$ is piecewise continuous, we need only show that $\mathbf{x}[\cdot]$ is piecewise continuous. But $\mathbf{x}[\cdot]$ solves $\dot{\mathbf{x}} = \mathbf{f}(\mathbf{x}, \mathbf{u}(t))$ a.e., and the right-hand side is piecewise continuous in t (since $\mathbf{f}$ is continuous and $\mathbf{u}(\cdot)$ is piecewise continuous), and so we can generate a piecewise continuous solution in the manner just outlined. □

3. Applying the Maximum Principle

EXAMPLE 3. We return to the Rocket Car, continuing the discussion of Section 5 of Chapter I:

$$\dot{p} = q, \qquad \dot{q} = u, \qquad \mathbf{x} = \begin{bmatrix} p \\ q \end{bmatrix},$$

with $C[u(\cdot)] = \int_{t_0}^{t_1} [q[s]]^2\, ds$ ($\lambda_1 = \lambda_3 = 0$, $\lambda_2 = 1$). We allow $u(\cdot)$ to be any *piecewise continuous* function satisfying $-1 \le u(t) \le 1$, ($u(\cdot) \in \mathscr{V}_{PCN}$) and we specify position zero, velocity zero as target: $\mathscr{T}(t) \equiv (0, 0)^T$. We will show that an optimal control exists only when the initial state $\mathbf{x}_0 = (p_0, q_0)^T$ lies on the "switching curve" $(+) \cup (-)$ sketched in Figure 5. To refresh the reader's memory, $(+)$ is the only response through the target when $u(t) \equiv +1$; $(-)$ is the only response through the target when $u(t) \equiv -1$. In Chapter II we gave a synthesis (with no cost restriction) which would take the system from any initial state to the target in the simplest manner (Figure 6):

(a) if $\mathbf{x}_0$ lies above $(+) \cup (-)$, use $u(t) \equiv -1$ until you hit $(+)$, then switch to $u(t) \equiv +1$.
(b) if $\mathbf{x}_0$ lies below $(+) \cup (-)$, use $u(t) \equiv +1$ until you hit $(-)$, then switch to $u(t) \equiv -1$.

Of course, this may not be the optimal way to do things when a cost function is involved. The PMP will tell us there is *no* optimal control unless we are on the curve $(+) \cup (-)$.

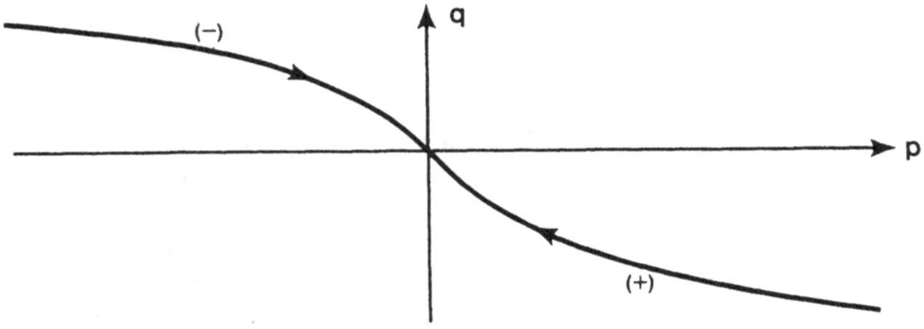

(+): $q=-\sqrt{2p},\quad p\geq 0$ (−): $q=\sqrt{-2p},\quad p\leq 0$

Figure 5

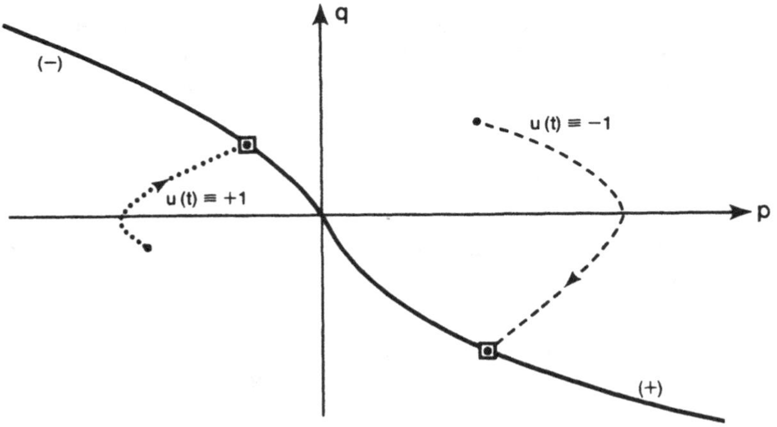

Figure 6 Dashed Trajectory Is Case (a); Dotted Trajectory Is Case (b); Switching State Is Labelled □

To apply the PMP, we set $t_0=0$ for notational convenience. Since the problem is autonomous, this involves no loss of generality. Our dynamic cost is $x^0[t]=\int_0^t [q[s]]^2\,ds$, and our extended state vector is $\hat{\mathbf{x}}[t]=(x^0[t], p[t], q[t])^T$. To keep superscripts to a minimum, we write

$$\text{(3)}\qquad \hat{\mathbf{w}}(t)=\begin{bmatrix}\alpha(t)\\ \beta(t)\\ \gamma(t)\end{bmatrix},\qquad \dot{\hat{\mathbf{w}}}(t)=-[\hat{\mathbf{f}}_{\hat{\mathbf{x}}}(\mathbf{x}[t], u(t))]^T\hat{\mathbf{w}}(t),$$

for any costate associated with a successful pair $(\mathbf{x}[\cdot], u(\cdot))$. Equation (3) can be written out by components:

$$\hat{\mathbf{f}}_{\hat{\mathbf{x}}}(\mathbf{x}, u)^T=\begin{bmatrix}0 & 0 & 0\\ 0 & 0 & 0\\ 2q & 1 & 0\end{bmatrix},\qquad \dot{\alpha}=0=\dot{\beta},\qquad \dot{\gamma}=-2q\alpha-\beta,$$

so $\alpha(t)\equiv\alpha(0)$, $\beta(t)\equiv\beta(0)$. Now

$$H(\hat{\mathbf{w}},\mathbf{x},v)=\langle\hat{\mathbf{w}},\hat{\mathbf{f}}(\mathbf{x},u)\rangle=\alpha[q]^2+\beta q+\gamma v.$$

The PMP asserts that if $(\hat{\mathbf{x}}[\cdot],u(\cdot))$ is optimal then there is a costate $\hat{\mathbf{w}}(\cdot)$ such that for each t in $[0,t_1]$, $H\equiv\alpha(0)[q[t]]^2+\beta(0)q[t]+\gamma(t)v\le 0$ for all numbers $-1\le v\le +1$, and $H=0$ for $v=u(t)$. In general, our equations $\dot{\hat{\mathbf{x}}}=\hat{\mathbf{f}}(\mathbf{x},u)$ and $(\widehat{\text{Adj.}})$ would only be satisfied a.e., but since $u(\cdot)$ is piecewise continuous, the corollary applies. By (iii) of the PMP,

$$w^0(t)\equiv\alpha(t)\equiv\alpha(0)\le 0.$$

Case 1. $\alpha(0)=0$. Then

$$\alpha(t)\equiv 0,\qquad \beta(t)\equiv\beta(0),\qquad \gamma(t)=-\beta(0)t+\gamma(0),$$

$$H=\langle\hat{\mathbf{w}},\hat{\mathbf{f}}\rangle=\beta(0)q[t]+[\gamma(0)-\beta(0)t]v,$$

$$\max_{-1\le v\le+1} H=M(\hat{\mathbf{w}},\mathbf{x})=\beta(0)q[t]+|\gamma(0)-\beta(0)t|\equiv 0.$$

We are using the control class $\mathscr{U}_{PC}$, so the corollary applies, and an optimal control must give $H=M=0$ *everywhere*. Therefore,

$$u(t)=\operatorname{sgn}[\gamma(0)-\beta(0)t],$$

$$M\equiv\beta(0)q[t]+|\gamma(0)-\beta(0)t|\equiv 0,\qquad 0\le t\le t_1.$$

Since $\alpha(0)=0$, by assumption, $\gamma(0)-\beta(0)t$ cannot vanish identically ($\hat{\mathbf{w}}(t)\ne\hat{\mathbf{0}}$). Thus $u(\cdot)$ is bang–bang with at most one switch.

If $\gamma(t)$ does not change sign, then either $u(t)\equiv+1$ or $u(t)\equiv-1$. In either case, we can only reach the target by starting on $(+)\cup(-)$.

Now suppose $\gamma(t)=\gamma(0)-\beta(0)t$ changes sign, say at $t=\tau>0$. Then $\gamma(0)\ne 0$ and $\beta(0)\ne 0$ and at τ,

$$\gamma(\tau)=0,\qquad H=\beta(0)q[\tau]=M=0.$$

Therefore $q[\tau]=0$. Now $u(t)$ made its single allowed switch at time τ, hence $u(t)\equiv+1$ or $u(t)\equiv-1$ for $t>\tau$. Therefore, to hit the target at some $t_1>\tau$, we must be on $(+)\cup(-)$ at time τ. But the only point on $(+)\cup(-)$ with $q[\tau]=0$ is the origin. Thus we had already arrived at the target at time τ, and switching was not necessary. But to have arrived at the target without switching, we must have originally started on $(+)\cup(-)$.

Case 2. $\alpha(t)\equiv\alpha(0)<0$. Since $(\widehat{\text{Adj.}})$ is linear, we can assume without loss of generality that $\alpha(t)\equiv-1$ (just multiply $\hat{\mathbf{w}}(t)$ by $-1/\alpha(0)$ to get a new solution to $(\widehat{\text{Adj.}})$ – this will not change max $H=0$).

$$H(\hat{\mathbf{w}},\mathbf{x},v)=-[q[t]]^2+\beta(0)q[t]+\gamma(t)v,$$

$$\max_{v\in\Omega} H=M=-[q[t]]^2+\beta(0)q[t]+|\gamma(t)|,$$

so as before $u(t) = \operatorname{sgn} \gamma(t)$. The function $\gamma(t)$ is in general no longer linear: $\dot{\gamma}(t) = 2q[t] - \beta(0)$. It is tempting to say that $u(t) = \operatorname{sgn} \gamma(t)$ is again bang–bang, but unfortunately $\gamma(t)$ might now vanish on large sets. We will show that $\gamma(t)$ can only vanish on at most a single interval (or at a single point). On such an interval, we will show that $u(t) \equiv 0$. Thus our optimal control may not be "bang–bang" but instead "bang–coast–bang." Finally, we will show that this latter case cannot happen, in fact we cannot have $u(t)$ switch from 0 or +1 to −1, nor can it switch from −1 or 0 to +1. Therefore *no* switching is possible, and $u(t) \equiv +1$ or $u(t) \equiv -1$. This forces us to start on $(+) \cup (-)$.

Now to the details. We have

$$\dot{\gamma}(t) = 2q[t] - \beta(0), \qquad \ddot{\gamma} = 2u(t) = 2 \operatorname{sgn} \gamma(t),$$

where $\ddot{\gamma}(\cdot)$ is piecewise continuous. Therefore

$$\gamma(\tau) > 0 \Rightarrow \gamma(t) \quad \text{concave up near } \tau,$$

$$\gamma(\tau) < 0 \Rightarrow \gamma(t) \quad \text{concave down near } \tau.$$

Figure 7 shows the eight possible local shapes for $\gamma(t)$; the dot indicates $(\tau, \gamma(\tau))$. Notice that $\gamma(t)$ *cannot* have two distinct zeros (try to sketch it). More precisely, if $\gamma(a) = \gamma(b) = 0$, then $\gamma(t) \equiv 0$ on $[a, b]$.

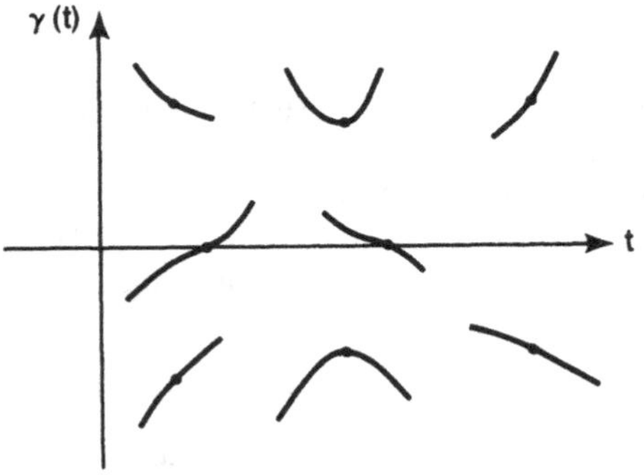

Figure 7

Therefore, $\gamma(t)$ can only vanish on a single interval. On such an interval I, $\gamma(t) \equiv 0$ which implies $\dot{\gamma}(t) \equiv 0$, therefore

$$2q[t] - \beta(0) \equiv 0 \Rightarrow q[t] \equiv \text{constant} \Rightarrow \dot{q}[t] \equiv 0.$$

But $\dot{q}[t] = u(t)$, so $u(t) \equiv 0$ on I.

Finally, we show that such a control cannot exist, in fact, no switching is possible.

Suppose $u(\cdot)$ switches from 0 or +1 to −1 at τ. Now $u(t) = \operatorname{sgn} \gamma(t)$ so the continuous function $\gamma(t)$ changes from non-negative to negative at τ, and (as remarked above) cannot vanish again. Thus $u(t) \equiv -1$ for $t \geq \tau$ and

$(p[\tau], q[\tau])$ must lie on $(+)\cup(-)$. Since $\dot{q}[t]=u(t)\equiv -1$ for $t\geq\tau$, and our target specifies $q[t_1]=0$, we must have $q[\tau]>0$. Now $\gamma(\tau)=0$, so at $t=\tau$,

$$H=-[q[\tau]]^2+\beta(0)q[\tau]=0.$$

But $q[\tau]>0$, so the above implies $q[\tau]=\beta(0)$. Since $\gamma(t)$ goes from non-negative to negative at τ,

$$0\geq\dot{\gamma}[\tau]=2q[\tau]-\beta(0)=q[\tau],$$

contradicting $q[\tau]>0$. Therefore no such switching point can exist for an optimal control. The argument for the case of switching from 0 or -1 to $+1$ is entirely analogous.

To summarize, we have shown that no switching points are allowed for the optimal control problem we are treating in this example. Therefore, an optimal control can only exist for initial states on $(+)\cup(-)$, and in this case either $u(t)\equiv +1$ or $u(t)\equiv -1$ for the entire time interval $[t_0, t_1]$.

EXAMPLE 4. We again consider the Rocket Car, but with a different cost:

$$C[u(\cdot)]=\frac{1}{2}\int_0^{t_1}(1+[q[t]]^2)\,dt \qquad (\lambda_1=\lambda_2=1/2, \lambda_3=0).$$

We will minimize $2C[u(\cdot)]$ for notational simplicity. The PMP gives us

$$(4)\qquad \begin{aligned}\dot{x}^0&=1+[q]^2\\ \dot{p}&=q,\\ \dot{q}&=u,\end{aligned}\qquad (\widehat{\text{Adj.}})\qquad \begin{aligned}\dot{\alpha}&=0,\\ \dot{\beta}&=0,\\ \dot{\gamma}&=-2q\alpha-\beta,\end{aligned}$$

with the Hamiltonian

$$H=\alpha(1+[q]^2)+\beta q+\gamma v.$$

$M(\hat{\mathbf{w}}, \mathbf{x})=\max_{v\in\Omega} H=\alpha(1+[q]^2)+\beta q+|\gamma|$, so

$$u(t)=\operatorname{sgn}\gamma(t).$$

Note that $\alpha(t)\equiv\alpha(0)$. By exactly the same argument as in the previous example, if $\alpha(t)\equiv\alpha(0)=0$, then $\mathbf{x}_0=(p_0, q_0)^T$ lies on $(+)\cup(-)$.

Now suppose that $\alpha(0)<0$; as before we may assume without loss of generality that $\alpha(t)\equiv\alpha(0)=-1$. In this case, we have

$$(5)\qquad \beta(t)\equiv\beta(0), \qquad \dot{\gamma}(t)=2q[t]-\beta(0),$$

and

$$H=-(1+[q]^2)+\beta(0)q[t]+\gamma(t)u(t).$$

As before, $\ddot{\gamma}=2u(t)=2\operatorname{sgn}\gamma(t)$, so $\gamma(t)$ can only have the local shapes sketched in Figure 7. Therefore, $\gamma(t)$ can vanish on at most a single interval (or point) I. Therefore, an optimal control $u(\cdot)$ is piecewise constant with

$$u(t)=\begin{cases}+1 & \text{when } \gamma(t)>0,\\ 0 & \text{when } \gamma(t)=0,\\ -1 & \text{when } \gamma(t)<0.\end{cases}$$

We will show that there is an optimal control for any initial state $\mathbf{x}_0$, and we shall synthesize it. Suppose $u(t)\equiv 0$ on an interval I, so $\gamma[t]\equiv 0$ on I. Then $\dot{\gamma}(t)\equiv 0$ and Equations (5) imply that on I, $2q[t]\equiv\beta(0)$. Also

$$H=-(1+[q[t]]^2)+\beta(0)q[t]\equiv 0,$$

so $[q[t]]^2\equiv 1$ on I. This means that if there is a time interval I on which $u(t)\equiv 0$, then we are letting the Rocket Car coast with fixed velocity $+1$ or -1.

We will show that the PMP selects a single candidate for an optimal control, for any initial state, and the synthesis is given by Figure 8. The fact that this control is in fact optimal follows from the discussion at the beginning of Section 6 of Chapter IV.

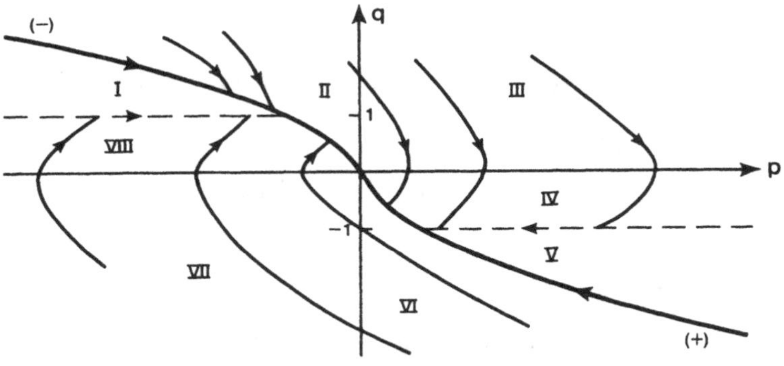

Figure 8 $u(t)\equiv 0$ on the Dashed Lines, $u(t)\equiv +1$ on (+), $u(t)\equiv -1$ on (−). $u(t)\equiv -1$ in Regions I, II, III, IV. $u(t)\equiv +1$ in Regions V, VI, VII, VIII

Some typical optimal responses are sketched in Figure 9. The control for $\mathbf{x}_0$ is "bang–off–bang" with $q_0<+1$. The control for $\mathbf{x}_0^{\#}$ is "bang–bang." The control for $\mathbf{x}_0^{*}$ is "bang–off–bang" with $q_0<-1$.

To show that the above synthesis is implied by the PMP, we first show that if $u(t)=\operatorname{sgn}\gamma(t)$ switches from 0 or $+1$ to -1 at time τ then $0<q[\tau]\leq 1$. At τ, $\gamma(\tau)=0$ and for $t\geq\tau$, $\ddot{\gamma}(t)=u(t)=\operatorname{sgn}\gamma(t)=-1$ since no more switches are allowed. Then $q[t]=u(t)=-1$ for $t\geq\tau$. Since the target state specifies $q[t_1]=0$, we must have $0\leq q[\tau]$. But $q[\tau]=0$ is impossible, since

$$H=-(1+[q[\tau]]^2)+\beta(0)q(\tau)=0. \tag{6}$$

Therefore $q[\tau]>0$, and (6) implies that both $\beta(0)$ and $q[\tau]$ are positive.

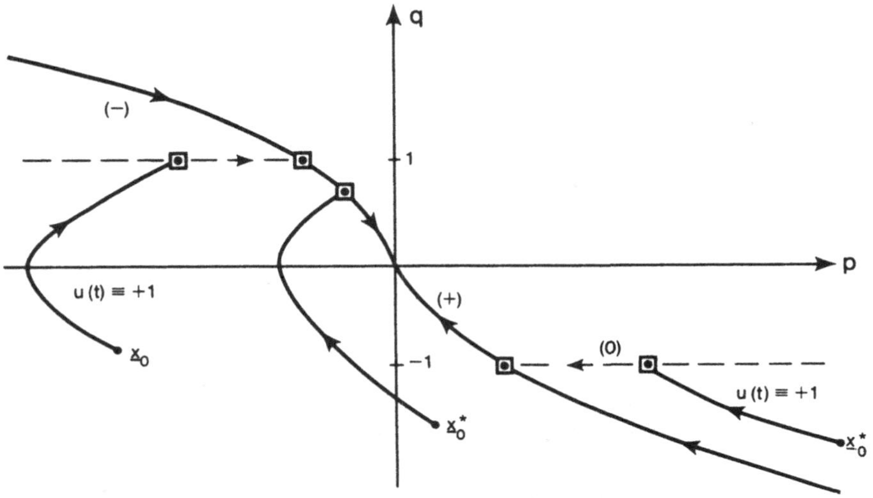

Figure 9

Since $\gamma(t)$ changed from non-negative to negative at τ,

$$0 \geq \dot{\gamma}(\tau) = 2q[\tau] - \beta(0).$$

Equation (6) and the above imply that

$$0 \geq -1 + [q[\tau]]^2, \quad \text{i.e.,} \quad 1 \geq [q[\tau]]^2.$$

Therefore $0 < q[\tau] \leq 1$. A dual argument shows that if $u(\cdot)$ switches from -1 or 0 to $+1$ at τ, then $-1 \leq q[\tau] < 0$.

The constraints on $q[\tau]$ at a switching time τ force us to switch if we hit the dashed lines in Figure 8 with a response that is moving away from the p-axis – we will never get another chance. A little experimentation shows that "bang–off–bang" is the *only* way to hit the origin for such a trajectory. (Remember we can only switch at most twice.) For trajectories that are moving toward the p-axis, the analysis is not difficult. From the state $\mathbf{x}_0$ in Figure 10, we can begin with $u(t) = +1$ (dashed), $u(t) = 0$ (dotted) or $u(t) = -1$ (solid). However, from our earlier analysis we know that the dotted line ("coasting") can only happen if $q[t] \equiv +1$, so the dotted line is out. The dashed trajectory is out because

(i) we cannot switch from $+1$ to -1 unless $-1 \leq q[\tau] < 0$. Therefore we can never switch to a "falling" trajectory.
(ii) Switching from $+1$ to 0 would require $q[\tau] = 1$ and we are always above $q = 1$.

Similar arguments cover the remaining cases.

A study of Figure 8 and the cost function $\int_0^{t_1} (1 + [q]^2)\, dt$ leads to the following remarks: for small velocities, the integral of $[q]^2$ is much smaller

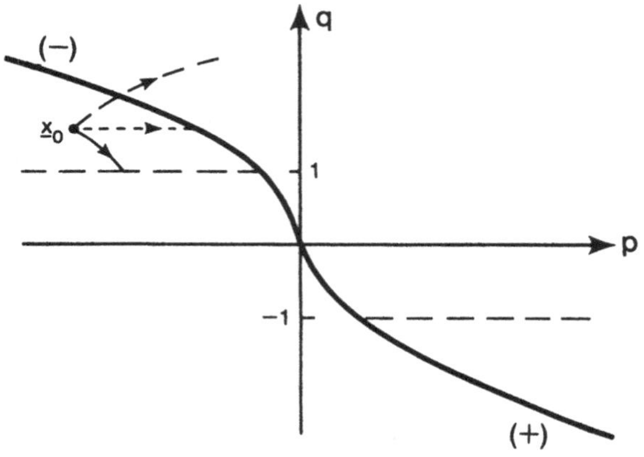

Figure 10 $\mathbf{x}_0 = (p_0, q_0)^T$ with $1 < q_0 < \sqrt{-p_0}$

than the integral of the constant term. Thus the strategy should be (and is) the time-optimal strategy. In fact, near the origin, the phase portrait is the same as the time-optimal phase portrait, given in Figure 7 of Chapter III. For large velocities, $[q]^2 \gg 1$, the strategy should be (and is) to decrease the velocity as quickly as possible. Similar analysis performed for the cost function

$$\lambda + (1-\lambda)[q]^2$$

suggests that as $\lambda \to 1$, the optimal phase portrait should "approach" that of Figure 7 of Chapter III. This is indeed the case.

4. A Dynamic Programming Approach to the Proof of the Maximum Principle

In this section we present a proof of the PMP which is valid under certain smoothness assumptions. Unfortunately, as we shall show by example, these assumptions often do not hold, but this proof shows how the costate arises in a natural way.

We consider the autonomous problem as stated in Section 2:

$$\dot{\mathbf{x}} = \mathbf{f}(\mathbf{x}, \mathbf{u}), \qquad C[\mathbf{u}(\cdot)] = \int_{t_0}^{t_1} f^0(\mathbf{x}[t], \mathbf{u}(t))\, dt$$

with t_0 and $\mathbf{x}(t_0) = \mathbf{x}_0$ specified. For this autonomous problem, we have a basic Principle of Optimality:

Any piece of an optimal trajectory is optimal.

More precisely (Figure 11), let $(\mathbf{u}(\cdot), \mathbf{x}[\cdot])$ be an optimal pair on $[t_0, t_1]$, steering state $\mathbf{x}_0$ to state $\mathbf{x}_1$.

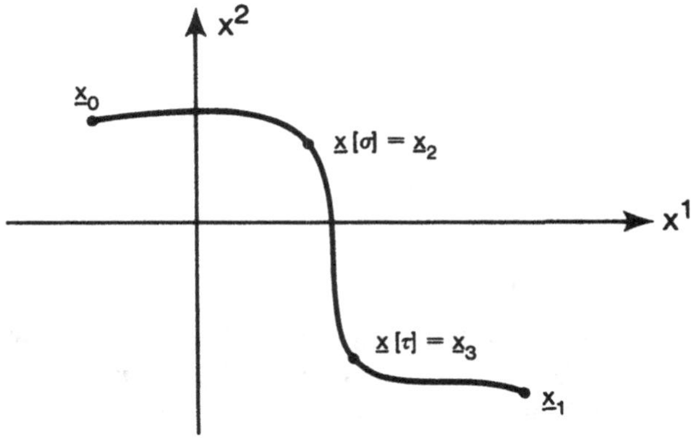

Figure 11 The Optimal Response $\mathbf{x}^*[t]$

If $[\sigma, \tau] \subset [t_0, t_1]$, then $\mathbf{u}(\cdot)$ is an optimal control steering state $\mathbf{x}[\sigma]$ to state $\mathbf{x}[\tau]$.

Proof. Suppose $\mathbf{u}^*(\cdot)$ is better than $\mathbf{u}(\cdot)$ for steering $\mathbf{x}_2 = \mathbf{x}[\sigma]$ to $\mathbf{x}_3 = \mathbf{x}[\tau]$. This implies that the solution $\mathbf{x}^*[t]$ of the problem

$$\dot{\mathbf{x}}^* = \mathbf{f}(\mathbf{x}^*, \mathbf{u}^*), \qquad \mathbf{x}^*(\sigma) = \mathbf{x}_2$$

eventually reaches the state $\mathbf{x}_3 = \mathbf{x}[\tau]$ at some time τ' with

$$\text{(7)} \qquad C[\mathbf{u}^*(\cdot)] = \int_\sigma^{\tau'} f^0(\mathbf{x}^*[s], \mathbf{u}^*(s))\, ds < \int_\sigma^{\tau} f^0(\mathbf{x}[s], \mathbf{u}(s))\, ds.$$

We can then piece together a better control than $\mathbf{u}(\cdot)$ for the original problem:

$$\mathbf{u}^{\#}(t) = \begin{cases} \mathbf{u}(t) & \text{on } [t_0, \sigma], \\ \mathbf{u}^*(t) & \text{on } (\sigma, \tau'), \\ \mathbf{u}(t - \tau' + \tau) & \text{on } [\tau', t_1 + \tau' - \tau], \end{cases}$$

with associated response:

$$\mathbf{x}^{\#}[t] = \begin{cases} \mathbf{x}[t] & \text{on } [t_0, \sigma], \\ \mathbf{x}^*[t] & \text{on } (\sigma, \tau'), \\ \mathbf{x}[t - \tau' + \tau] & \text{on } [\tau', t_1 + \tau' - \tau]. \end{cases}$$

Because $\mathbf{f}$ and f^0 are independent of t, this response satisfies $\dot{\mathbf{x}}^{\#} = \mathbf{f}(\mathbf{x}^{\#}, \mathbf{u}^{\#})$ a.e. on $[t_0, t_1 + \tau' - \tau]$, and $\mathbf{x}^{\#}[t_0] = \mathbf{x}_0$, $\mathbf{x}^{\#}[t_1 + \tau' - \tau] = \mathbf{x}[t_1 + \tau' - \tau - \tau' + \tau] =$

$\mathbf{x}[t_1]=\mathbf{x}_1$. The cost is lower than the cost for $\mathbf{u}(\cdot)$:

$$\begin{aligned} C[\mathbf{u}^{\#}(\cdot)] &= \int_{t_0}^{t_1+\tau'-\tau} f^0(\mathbf{x}^{\#}[s], \mathbf{u}^{\#}(s))\, ds \\ &= \int_{t_0}^{\sigma} f^0(\mathbf{x}[s], \mathbf{u}(s))\, ds \\ &\quad + \int_{\sigma}^{\tau'} f^0(\mathbf{x}^*[s], \mathbf{u}^*(s))\, ds \\ &\quad + \int_{\tau'}^{t_1+\tau'-\tau} f^0(\mathbf{x}[s-\tau'+\tau], \mathbf{u}(s-\tau'+\tau))\, ds. \end{aligned}$$

A change of variable in the last term reduces it to $\int_\tau^{t_1} f^0(\mathbf{x}[r], \mathbf{u}[r])\, dr$. The middle term on the right is strictly less than $\int_\sigma^\tau f^0(\mathbf{x}[s], \mathbf{u}(s))\, ds$, by (7). Thus $C[\mathbf{u}^{\#}(\cdot)] < C[\mathbf{u}(\cdot)]$, contradicting the optimality of $\mathbf{u}(\cdot)$. □

The above Optimality Principle is useful in a wide variety of situations. We will use it to prove:

Theorem 2. *Consider the optimal control problem stated at the beginning of Section 2. If for each initial state $\mathbf{x}_0$ in some neighborhood N of the target state $\mathbf{x}_1$ there is an optimal control $\mathbf{u}^*(t, \mathbf{x}_0)$ with associated response $\mathbf{x}^*[t]$ which stays in N, and if the associated minimum cost $C[\mathbf{u}^*(t, \mathbf{x}_0)]$ is twice continuously differentiable with respect to $\mathbf{x}_0$, then for each $\mathbf{x}_0 \in N$ there is an extended costate $\hat{\mathbf{w}}(\cdot)$ satisfying ($\widehat{\text{Adj}}$.) a.e., for which $H(\hat{\mathbf{w}}(t), \mathbf{x}^*[t], \mathbf{v}) \le 0$ for all $\mathbf{v} \in \Psi$ and all $t \in [0, t_1]$; $H(\hat{\mathbf{w}}(t), \mathbf{x}^*[t], \mathbf{u}^*(t, \mathbf{x}_0)) = 0$ a.e. in t.*

Proof. Let $\mathbf{u}^*(t, \mathbf{x}_0)$ *always denote the optimal control from* $\mathbf{x}_0 \in N$. For simplicity we write $C(\mathbf{x}_0)$ for $C[\mathbf{u}^*(t, \mathbf{x}_0)]$, the lowest cost in getting from state $\mathbf{x}_0$ to the target state $\mathbf{x}_1$. We choose an instant σ near time 0, $\sigma > 0$, and a constant vector $\mathbf{v} \in \Psi$, to form the perturbed control

$$\tilde{\mathbf{u}}(t, \mathbf{x}_0) = \begin{cases} \mathbf{v}, & 0 \le t \le \sigma, \\ \mathbf{u}^*(t, \mathbf{x}_0), & \sigma < t \le t_1. \end{cases}$$

The associated response is denoted $\tilde{\mathbf{x}}[\cdot]$ (Figure 12). It is possible (in fact probable) that $\mathbf{v}$ will *not* be the optimal control for getting from state $\mathbf{x}_0$ to the state $\tilde{\mathbf{x}}[\sigma]$. In fact, for any t between 0 and σ,

$$C(\mathbf{x}_0) \le \int_0^t f^0(\tilde{\mathbf{x}}[s], \mathbf{v})\, ds + C(\tilde{\mathbf{x}}[t]).$$

We rewrite this as

$$C(\tilde{\mathbf{x}}[0]) - C(\tilde{\mathbf{x}}[t]) \le \int_0^t f^0(\tilde{\mathbf{x}}[s], \mathbf{v})\, ds.$$

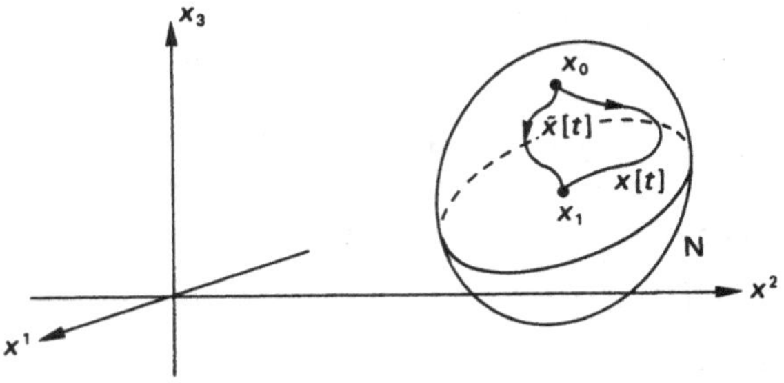

Figure 12 $(\mathbf{u}^*(\cdot), \mathbf{x}^*[\cdot])$ Is Optimal

Note that $\mathbf{x}[\cdot]$ is continuously differentiable on $[0, \sigma)$, since it solves $\dot{\tilde{\mathbf{x}}} = \mathbf{f}(\tilde{\mathbf{x}}, \mathbf{v})$ on this interval. Therefore, $f^0(\tilde{\mathbf{x}}[s], \mathbf{v})$ is continuous in s, so dividing the above by t and letting $t \to 0$, we get

$$-\frac{d}{dt} C(\tilde{\mathbf{x}}[t])\bigg|_{t=0} \leq f^0(\mathbf{x}_0, \mathbf{v}) \quad \text{for all } \mathbf{v} \in \Psi.$$

Under our smoothness assumptions we can write this in the form

$$\text{(8)} \qquad -\langle \text{grad}_{\mathbf{x}}\, C(\mathbf{x}), \mathbf{f}(\mathbf{x}, \mathbf{v})\rangle|_{\mathbf{x}=\mathbf{x}_0} \leq f^0(\mathbf{x}_0, \mathbf{v}),$$

for all $\mathbf{v} \in \Psi$. But $\mathbf{x}_0 \in N$ was arbitrary, hence this last equation holds with $\mathbf{x}_0$ replaced by any $\mathbf{x} \in N$.

Given a fixed $\mathbf{x}_0 \in N$, for the optimal pair $(\mathbf{u}^*(t, \mathbf{x}_0), \mathbf{x}^*[t])$ we define an associated $(n+1)$-vector $\hat{\mathbf{w}}(t)$ as follows:

$$\hat{\mathbf{w}}(t) = \begin{bmatrix} -1 \\ -\text{grad}_{\mathbf{x}} \quad C(\mathbf{x}) \end{bmatrix}\Bigg|_{\mathbf{x}=\mathbf{x}^*[t]} = -\text{grad}_{\hat{\mathbf{x}}}\, (x^0, C(\mathbf{x}))|_{\mathbf{x}=\mathbf{x}^*(t)}.$$

Then (8) is equivalent to

$$\langle \hat{\mathbf{w}}(t), \hat{\mathbf{f}}(\mathbf{x}^*[t], \mathbf{v})\rangle \leq 0, \qquad \hat{\mathbf{f}} = (f^0, \mathbf{f}^T)^T.$$

This is just the statement $H(\hat{\mathbf{w}}(t), \mathbf{x}^*[t], \mathbf{v}) \leq 0$. To complete the proof, we must show that

(1) $H(\hat{\mathbf{w}}(t), \mathbf{x}^*[t], \mathbf{u}^*(t)) = 0$ a.e.,
(2) $\hat{\mathbf{w}}(t)$ solves $(\widehat{\text{Adj.}})$ a.e.

To prove (1) we note that the Principle of Optimality implies that if $\mathbf{u}^*(\cdot)$ optimally steers $\mathbf{x}_0$ to $\mathbf{x}_1$, with response $\mathbf{x}^*[\cdot]$, then

$$C(\mathbf{x}_0) = \int_{t_0}^{t} f^0(\mathbf{x}^*[s], \mathbf{u}^*(s))\, ds + C(\mathbf{x}^*[t]),$$

for $t_0 < t < t_1$. We can differentiate to obtain

$$0 = f^0(\mathbf{x}^*[t], \mathbf{u}^*(t)) + \langle \operatorname{grad}_{\mathbf{x}} C(\mathbf{x}), \mathbf{f}(\mathbf{x}, \mathbf{u}^*(t))\rangle|_{\mathbf{x}=\mathbf{x}^*[t]}$$

a.e. in t. Therefore (1) holds.

To prove (2), we first note that if t is fixed, then

$$H(\hat{\mathbf{w}}, \mathbf{x}, \mathbf{u}) = -f^0(\mathbf{x}, \mathbf{u}) - \langle \operatorname{grad}_{\mathbf{x}} C(\mathbf{x}), \mathbf{f}(\mathbf{x}, \mathbf{u})\rangle$$

is continuously differentiable with respect to $\mathbf{x}$ in N, and attains a maximum whenever we substitute $(\mathbf{u}^*(t), \mathbf{x}^*[t])$ for any t in $[t_0, t_1]$ (a.e.). Therefore, if we fix both t and $\mathbf{u} = \mathbf{u}^*(t)$, and let $\mathbf{x}$ vary, then

$$\operatorname{grad}_{\mathbf{x}} H(\hat{\mathbf{w}}(t), \mathbf{x}, \mathbf{u}^*(t)) = \mathbf{0} \quad \text{at } \mathbf{x} = \mathbf{x}^*[t], \quad \text{a.e. in } t.$$

Now

$$\operatorname{grad}_{\mathbf{x}} H(\hat{\mathbf{w}}(t), \mathbf{x}, \mathbf{u}^*(t)) = -\operatorname{grad}_{\mathbf{x}} f^0(\mathbf{x}, \mathbf{u}^*(t)) - \operatorname{grad}_{\mathbf{x}} \sum_{j=1}^{n} \frac{\partial C}{\partial x^j} f^j(\mathbf{x}, \mathbf{u}^*(t)),$$

so $\operatorname{grad}_{\mathbf{x}} H(\hat{\mathbf{w}}(t), \mathbf{x}, \mathbf{u}^*(t))|_{\mathbf{x}=\mathbf{x}^*[t]} = \mathbf{0}$ implies

$$\frac{\partial f^0}{\partial x^i}(\mathbf{x}^*[t], \mathbf{u}^*(t)) + \sum_{j=1}^{n} \frac{\partial^2 C(\mathbf{x}^*[t])}{\partial x^i \partial x^j} f^j(\mathbf{x}^*[t], \mathbf{u}^*(t)) + \sum_{j=1}^{n} \frac{\partial C(\mathbf{x}^*[t])}{\partial x^j} \frac{\partial f^j(\mathbf{x}^*[t], \mathbf{u}^*(t))}{\partial x^i} = 0, \qquad i = 1, 2, \ldots, n,$$

a.e. in t. With

$$\hat{\mathbf{w}}(t) = \begin{bmatrix} -1 \\ -\operatorname{grad}_{\mathbf{x}} C(\mathbf{x}^*[t]) \end{bmatrix}$$

this last implies that $\hat{\mathbf{w}}(\cdot)$ satisfies

$$\dot{w}^0(t) = 0 \quad (\text{in fact } w^0(t) \equiv -1),$$

$$\dot{w}^k(t) = \left\langle -\mathbf{w}, \frac{\partial \mathbf{f}}{\partial x^k} \right\rangle + \frac{\partial f^0}{\partial x^k} = -\left\langle \hat{\mathbf{w}}, \frac{\partial \hat{\mathbf{f}}}{\partial x^k} \right\rangle \quad \text{a.e.}$$

for $k = 1, \ldots, n$, i.e., $\dot{\hat{\mathbf{w}}} = -[\hat{\mathbf{f}}_{\hat{\mathbf{x}}}(\mathbf{x}^*[t], \mathbf{u}^*(t))]^T \hat{\mathbf{w}}$ a.e. □

Remarks

(1) Notice that $w^0(t) \neq 0$ for all t, and that $\mathbf{w}(t) = -\operatorname{grad}_{\mathbf{x}} C(\mathbf{x})$ is normal to the surfaces of constant cost, $C(\mathbf{x}) = \text{constant}$.

(2) The hypothesis that $C(\mathbf{x})$ be twice differentiable is quite severe; as will be shown in Example 5 below, this hypothesis is false even for the

Rocket Car. One can eliminate this hypothesis by assuming some rather complicated conditions involving the dimension of the set where $C(\mathbf{x})$ is not smooth. For the details, see Fleming and Rishel [1975], Chapter IV, Section 6.

EXAMPLE 5 ($C(\mathbf{x})$ is not always differentiable). As we showed intuitively in Section 5 of Chapter I, and verified in Examples 1 and 7 of Chapter III and Example 2 of this chapter, the Rocket Car problem has an unique time-optimal bang–bang control of the form $u(t) = \text{sgn}(\alpha - \beta t)$. In fact (Figure 13)

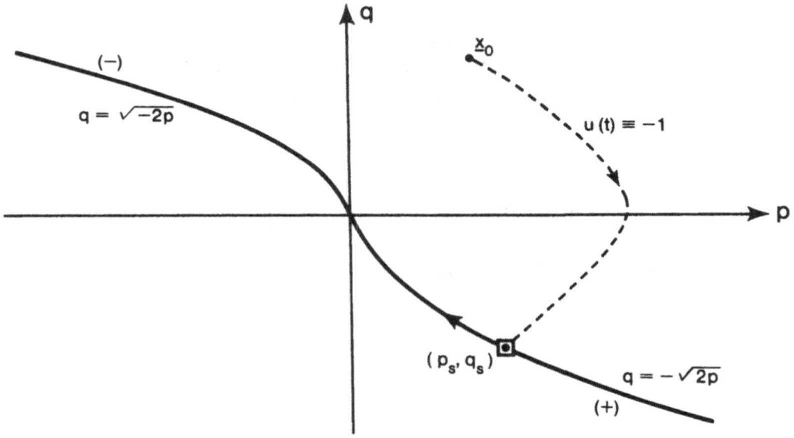

Figure 13

We will first compute $C(\mathbf{x}_0)$ for $\mathbf{x}_0$ above $(+)\cup(-)$, taking $t_0 = 0$. For $u(t) \equiv -1$, we can solve the equations of motion $\dot{p} = q$, $\dot{q} = u$ to get $[q[t]]^2 = q_0^2 + 2p_0 - 2p[t]$. The switching point (p_s, q_s) occurs where this curve intersects $q^2 = 2p$, which gives $(p_s, q_s) = (\frac{1}{4}(2p_0 + q_0^2), (p_0 + \frac{1}{2}q_0^2)^{1/2})$, and the switching time is $t_s = q_0 - q_s$. After the switch, we can again solve the equations of motion to conclude that the elapsed time from switch to arrival at target is $t_1 - t_s = t_s - q_s = q_0 - 2q_s$ (note $q_s < 0$). Therefore, for the time-optimal problem, with $\mathbf{x}_0$ above $(+)\cup(-)$,

$$C[u(\cdot)] = t_1 = q_0 + (2[q_0]^2 + 4p_0)^{1/2}.$$

The same type of analysis for $\mathbf{x}_0$ below $(+)\cup(-)$ gives $t_1 = -q_0 + \sqrt{2[q_0]^2 - 4p_0}$. Therefore

$$C(\mathbf{x}_0) = \begin{cases} q_0 + (2[q_0]^2 + 4p_0)^{1/2} & \text{above } (+)\cup(-), \\ -q_0 + (2[q_0]^2 - 4p_0)^{1/2} & \text{below } (+)\cup(-). \end{cases}$$

This function is continuous but not differentiable on $(+)\cup(-)$.

5. The PMP for More Complicated Problems

We will treat two extensions of the PMP. The first is for the case when the initial and target states $\mathbf{x}[t_0]=\mathbf{x}_0$, $\mathbf{x}[t_1]=\mathbf{x}_1$ are replaced by *sets* S_0, S_1. The second is to non-autonomous problems – this extension will make use of the first.

Let S_0 and S_1 be smooth manifolds, of dimension $1\le r_0<n$, $1\le r_1<n$ respectively. The simplest way to define a smooth r-dimensional manifold in R^n is as the intersection of surfaces defined by d implicit equations:

$$g^i(x^1, x^2, \ldots, x^n)=0, \qquad i=1,\ldots,d$$

with $1\le d\le n$ and $\operatorname{rank}[\partial g^i/\partial x^j]=n-r$ on an open set in R^n containing the manifold. For example, the vertical line $L_0=\{(0,0,\xi)|\xi\in R\}$ can be thought of as the intersection of the two planes $g^1(x^1,x^2,x^3)\equiv x^1=0$ and $g^2(x^1,x^2,x^3)\equiv x^2=0$ (Figure 14). Here,

$$\operatorname{rank}[\partial g^i/\partial x_j]=\operatorname{rank}\begin{bmatrix}1 & 0\\ 0 & 1\end{bmatrix}=2,$$

so it is a one-dimensional manifold.

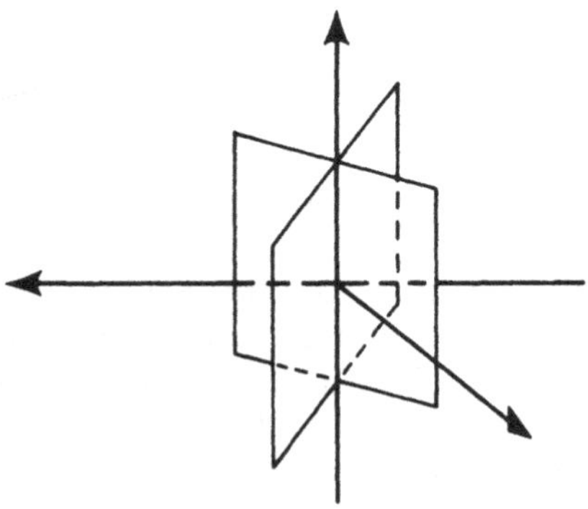

Figure 14

Theorem 3. *Consider the problem*

$$\dot{\mathbf{x}}=\mathbf{f}(\mathbf{x},\mathbf{u}), \qquad \mathbf{x}[t_0]\in S_0, \qquad \mathbf{x}[t_1]\in S_1,$$

with cost $C(\mathbf{u}(\cdot))=\int_{t_0}^{t_1} f^0(\mathbf{x}[s],\mathbf{u}(s))\,ds$. *If* $(\mathbf{u}^*(\cdot),\mathbf{x}^*[\cdot])$ *is an optimal pair, then there exists a continuous* $(n+1)$*-vector function* $\hat{\mathbf{w}}(\cdot)$ *solving* $(\widehat{Adj.})$ *a.e. and satisfying conclusions* (i), (ii) *and* (iii) *of the PMP. Also, if* T_0 *and* T_1 *are the tangent planes to* S_0 *and* S_1 *at* $\mathbf{x}[t_0]$, $\mathbf{x}[t_1]$ *respectively, then we can select the vector function* $\hat{\mathbf{w}}(t)$ *so that* $\mathbf{w}(t)$ *is perpendicular to* T_0 *at time* t_0

and perpendicular to T_1 at time t_1:

(9) $$\langle \mathbf{w}(t_0), \mathbf{v}_0\rangle = 0 = \langle \mathbf{w}(t_1), \mathbf{v}_1\rangle \quad \forall \mathbf{v}_0 \in T_0, \quad \mathbf{v}_1 \in T_1.$$

Remarks. The conditions (9) are called the *transversality conditions*; they assert that the vector $\mathbf{w}(t)$ (*not* the curve described by $\mathbf{w}(\cdot)$) is perpendicular to S_0 and S_1 at times t_0 and t_1 respectively. For a proof of this theorem, see Lee and Markus [1967], Chapter 5.

EXAMPLE 6. Consider the time-optimal problem for the Rocket Car:

$$\dot{p} = q, \qquad \dot{q} = u, \qquad C[u(\cdot)] = t_1 - t_0 = \int_{t_0}^{t_1} ds,$$

with piecewise continuous controls satisfying $-1 \le u(t) \le 1$. Suppose that $\mathbf{x}_0 = (p_0, q_0)^T$ is fixed, and the q-axis is the target:

$$S_0 = \{\mathbf{x}_0\}, \qquad S_1 = \{(0, \xi) | \xi \text{ real}\}.$$

This means we want to steer to the zero position, $p = 0$, in minimum time without worrying about our terminal velocity. S_0 is 0-dimensional, so there is no transversality condition at $\mathbf{x}[t_0]$. Theorem 3 asserts that:

$$\begin{aligned} &\dot{x}^0 = f^0(\mathbf{x}, u) = 1 && \dot{\alpha} = 0, && \\ &\dot{p} = q, && (\widehat{\text{Adj.}})\ \dot{\beta} = 0, && \hat{\mathbf{w}} = \begin{bmatrix} \alpha \\ \beta \\ \gamma \end{bmatrix}; \\ &\dot{q} = u, && \dot{\gamma} = -\beta, && \\ &H = \alpha + \beta q + \gamma v; && \langle \mathbf{w}(t_1), \mathbf{b}\rangle = 0, && \mathbf{w} = \begin{bmatrix} \beta \\ \gamma \end{bmatrix}, \end{aligned}$$

for any vector $\mathbf{b}$ tangent to S_1. Clearly such a vector has the form $\mathbf{b} = (0, \xi)^T$, so the transversality condition becomes $0 = \langle \mathbf{w}(t), \mathbf{b}\rangle = \gamma(t_1)\xi$. Thus $\gamma(t_1) = 0$. As in earlier examples, we can easily show that $\alpha(t) \equiv \alpha(0)$, $\beta(t) \equiv \beta(0)$, $\gamma(t) = \gamma(0) - \beta(0)t$ and $u(t) = \operatorname{sgn} \gamma(t)$.

We claim $\gamma(t) \ne 0$ on $[0, t_1)$. If $\gamma(\cdot)$ had another zero, then $\dot{\gamma}(t_\#) = 0$ for some $t_\#$. Then, since $\dot{\gamma} = -\beta$, we would have $\beta(t) \equiv \beta(0) = 0$. Then at t_1, we would have $H = \alpha(0)$. But $\max_{v \in \Omega} H = 0$ would imply $\alpha(0) = 0$, in which case $\hat{\mathbf{w}}(t_1) = \hat{\mathbf{0}}$ contradicting the PMP.

Therefore, $\gamma(t) \ne 0$ on $[t_0, t_1)$. Thus $u(t) = \operatorname{sgn} \gamma(t)$ is either identically $+1$ or -1 on $[t_0, t_1)$. Theorem 3 has given us the intuitively obvious solution "full power until you hit $p = 0$."

There is a hitch, however. Referring to Figure 15, in state space regions Q_1 and Q_3 we obviously use $u \equiv +1$, $u \equiv -1$ respectively. But in the shaded regions Q_2 and Q_4, we can hit the q-axis ($p = 0$) using either $u \equiv +1$ or $u \equiv -1$ (recall $\dot{q} = u$, $\dot{p} = q$); we have sketched the two possible trajectories as dashed curves for one state $\mathbf{x}_0$:

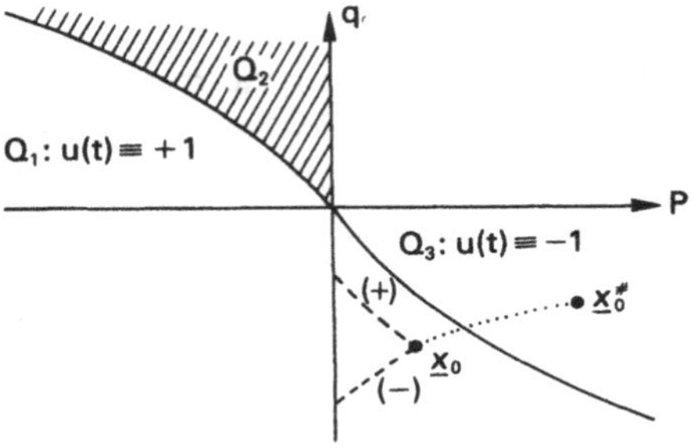

Figure 15

Here we can use the Principle of Optimality. Beginning in the state $\mathbf{x}_0^{\#}$ in Q_3, we use the optimal control $u \equiv -1$, generating the dotted trajectory through $\mathbf{x}_0$. By the Principle of Optimality, for each initial state on this trajectory, we use $u(t) \equiv -1$. Since the region Q_4 can be covered by such optimal trajectories (by varying $\mathbf{x}_0^{\#}$), we conclude that $u(t) \equiv -1$ is always optimal for $p_0 > 0$. A dual argument shows $u(t) \equiv +1$ is optimal for $p_0 < 0$.

We now turn to the PMP for non-autonomous problems. We consider the problem

$$\text{(NA)} \qquad \dot{\mathbf{x}} = \mathbf{f}(t, \mathbf{x}, \mathbf{u}), \qquad C(\mathbf{u}(\cdot)) = \int_{t_0}^{t_1} f^0(t, \mathbf{x}[t], \mathbf{u}(t))\, dt$$

with $\mathbf{x}_0$ and $\mathcal{T}(t) \equiv \mathbf{x}_1$ specified. We will use measurable controls taking values in a fixed bounded set. *We assume that t_0 is fixed*, and as usual allow t_1 to vary. We define a new coordinate x^{n+1} by

$$\dot{x}^{n+1} = 1, \qquad x^{n+1}[t_0] = t_0,$$

and we adjoin this initial value problem to (NA) to get the autonomous $(n+1)$-dimensional problem:

$$\text{(10)} \quad \begin{Bmatrix} \dot{\mathbf{x}}[t] = \mathbf{f}(x^{n+1}, \mathbf{x}, \mathbf{u}) \\ \dot{x}^{n+1}[t] = 1 \end{Bmatrix}, \qquad C(\mathbf{u}(\cdot)) = \int_{t_0}^{t_1} f^0(x^{n+1}[t], \mathbf{x}[t], \mathbf{u}(t))\, dt,$$

with initial state $(\mathbf{x}_0, t_0)$ and as target the *line*:

$$\mathcal{T}^*(t) = \{(\mathbf{x}_1, t_1) | t_1 \geq t_0\}.$$

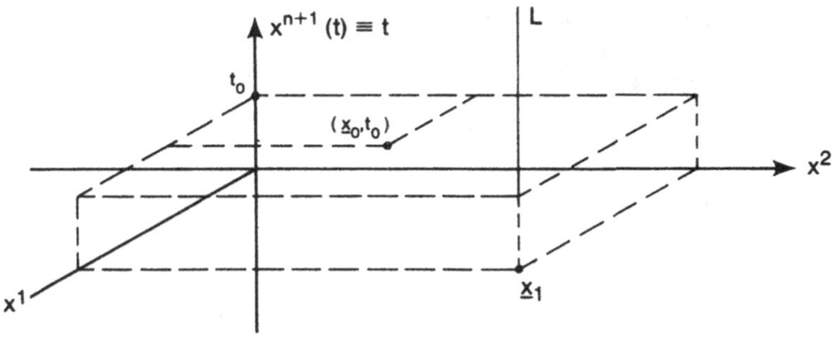

Figure 16

We can now use the autonomous version of the PMP with transversality to find necessary conditions for optimal controls and responses, in terms of the $(n+2)$-dimensional vector $(x^0, \mathbf{x}, x^{n+1})$, where as always

$$\dot{x}^0 = f^0(t, \mathbf{x}, \mathbf{u}) = f^0(x^{n+1}, \mathbf{x}, \mathbf{u}).$$

Using the fact that $x^{n+1}[t] \equiv t$, we may then rewrite these conditions in terms of the usual extended state vector $\hat{\mathbf{x}} = (x^0, \mathbf{x})$ and the usual Hamiltonian H as follows:

Theorem 4. *Assume that* $\mathbf{f}(t, \mathbf{x}, \mathbf{u})$ *and* $f^0(t, \mathbf{x}, \mathbf{u})$ *are continuous in* $(t, \mathbf{x}, \mathbf{u})$ *and continuously differentiable with respect to* $(t, \mathbf{x})$. *If* $(\mathbf{u}(\cdot), \mathbf{x}[\cdot])$ *is an optimal pair for* (*NA*), *then there exists a non-zero absolutely continuous* $(n+1)$*-vector function* $\hat{\mathbf{w}}(\cdot)$ *such that on* $[t_0, t_1]$ *with* $H = \langle \hat{\mathbf{w}}, \hat{\mathbf{f}} \rangle$,

$$w^0(t) \equiv w^0(t_0) \leq 0,$$

$$\dot{\hat{\mathbf{w}}}(t) = -[\hat{\mathbf{f}}_{\hat{\mathbf{x}}}(t, \mathbf{x}[t], \mathbf{u}(t))]^T \hat{\mathbf{w}}(t) \quad \text{a.e.},$$

$$H(\hat{\mathbf{w}}(t), \mathbf{x}[t], t, \mathbf{u}(t)) = M(\hat{\mathbf{w}}(t), \mathbf{x}[t], t) \quad \text{a.e.},$$

$$M(\hat{\mathbf{w}}(t), \mathbf{x}[t], t) \equiv \int_{t_0}^{t} \sum_{k=0}^{n} w^k(s) \frac{\partial}{\partial s} f^k(s, \mathbf{x}[s], \mathbf{u}(s))\, ds + K \text{ (a constant)}.$$

As always, $M = \max_{\mathbf{v} \in \Psi} H(\hat{\mathbf{w}}(t), \mathbf{x}[t], t, \mathbf{v})$, and $H = \langle \hat{\mathbf{w}}, \hat{\mathbf{f}} \rangle$. The analogues of Lemma 1 and its corollary also hold for (NA).

Example 7. In the appendix to Chapter III, we proved the autonomous version of the Bang–Bang Principle. Theorem 4 allows us to prove the time dependent version.

Let $\mathbf{u}(\cdot)$ be an optimal control on $[0, t_1]$ for the problem: $\dot{\mathbf{x}} = A(t)\mathbf{x} + B(t)\mathbf{u} + \mathbf{c}(t)$, with $C[\mathbf{u}(\cdot)] = t_1 - t_0$, and measurable controls with values in the unit cube Ω. Then there is a non-zero vector $\mathbf{b} \in R^n$ such that

$$u^i(t) = \operatorname{sgn}\{\mathbf{b}^T X^{-1}(t) B(t)\}^i, \qquad i = 1, 2, \ldots, m,$$

where $X(t)$ is a fundamental matrix for $\dot{\mathbf{x}} = A(t)\mathbf{x}$.

The associated systems are

$$(11)\quad \begin{array}{ll}\text{(i)} & \dot{x}^0 = f^0 = 1\\ \text{(ii)} & \dot{\mathbf{x}} = A(t)\mathbf{x} + B(t)\mathbf{u} + \mathbf{c}(t)\end{array} \qquad (\widehat{\text{Adj.}})\quad \begin{array}{ll}\text{(i)} & \dot{w}^0 = 0\\ \text{(ii)} & \dot{\mathbf{w}} = -A^T(t)\mathbf{w}\end{array}$$

with

$$H = \langle \hat{\mathbf{w}}, \hat{\mathbf{f}} \rangle = w^0 + \mathbf{w}^T A(t)\mathbf{x} + \mathbf{w}^T B(t)\mathbf{u} + \mathbf{w}^T \mathbf{c}.$$

Clearly, $M = \max_{\mathbf{v}\in\Psi} H = \mathbf{w}^0 + \mathbf{w}^T A(t)\mathbf{x} + \mathbf{w}^T \mathbf{c} + \max_{\mathbf{v}\in\Psi} \mathbf{w}^T B(t)\mathbf{v}$. Now if $X(t)$ is a fundamental matrix for $\dot{\mathbf{x}} = A(t)\mathbf{x}$ then $[X^{-1}(t)]^T$ is a fundamental matrix for $(\widehat{\text{Adj.}})$ (ii). Since $\mathbf{w}(\cdot)$ solves $(\widehat{\text{Adj.}})$ (ii), it must be of the form $\mathbf{w}(t) = [X^{-1}(t)]^T \mathbf{b}$ for some constant vector $\mathbf{b}$. Then by Theorem 4, the optimal control $\mathbf{u}(\cdot)$ maximizes

$$\mathbf{w}^T B(t)\mathbf{v} = \mathbf{b}^T X^{-1}(t) B(t)\mathbf{v}, \qquad -1 \le v^i \le 1.$$

Clearly, this last is maximized for

$$u^i(t) = v^i = \operatorname{sgn}\{\mathbf{b}^T X^{-1}(t) B(t)\}^i \qquad i = 1, \ldots, m.$$

Exercises

Remark: Unless stated otherwise, the time of arrival at the target is not specified.

1. For a given initial value problem in R^n,

$$\dot{\mathbf{x}} = \mathbf{g}(\mathbf{x}), \qquad \mathbf{x}(t_0) = \mathbf{x}_0,$$

let $\mathbf{x}(t)$ be a fixed solution. Consider the associated linearized system

$$\dot{\mathbf{b}} = \mathbf{g}_{\mathbf{x}}(\mathbf{x}(t))\mathbf{b}$$

Show that if $\mathbf{b}_0$ is tangent to the curve $\mathbf{x}(\cdot)$ at $\mathbf{x}(t_0)$, then the solution $\mathbf{b}(t)$ that satisfies $\mathbf{b}(t_0) = \mathbf{b}_0$ will be tangent to $\mathbf{x}(\cdot)$ for all t, $t_0 \le t \le t_1$.

2. If $\mathbf{b}(\cdot)$ solves (Lin.) and $\mathbf{w}(\cdot)$ solves (Adj.), show that

$$\langle \mathbf{w}(t), \mathbf{b}(t) \rangle \equiv \text{constant}.$$

3. In Example 3 of this chapter, we showed that the Rocket Car problem:

$$\dot{p} = q, \qquad \dot{q} = u, \qquad C[u(\cdot)] = \int_{t_0}^{t} [q(s)]^2\, ds$$

has no solution if (p_0, q_0) is off $(+)\cup(-)$. Which conditions fail from Theorem 3 of Chapter IV? (Cf. the discussion in Section 6 of Chapter IV.)

4. (Continuing Example 2.) Consider the Rocket Car: $\dot{p} = q$, $\dot{q} = u$, with $C[u(\cdot)] = t_1$, $\mathscr{T}(t) = \mathbf{0}$, and $u(\cdot)$ piecewise smooth *with* $0 \le u(t) \le 2$. Describe the controllable set, prove that an optimal control exists for all $\mathbf{x}_0 \in \mathscr{C}$, and use the PMP to synthesize the optimal control.

5. (Continuing Exercise 1 from Chapter III.) Consider the scalar problem $\dot{x} = u$ with $x_0 = -1$, $\mathscr{T}(t) \equiv 0$, $-1 \le u(t) \le +1$ with $u(\cdot)$ piecewise continuous. If $C[u(\cdot)] = t_1$, form H and (Adj.), and synthesize the optimal control.

6. Consider the linear system $\dot{\mathbf{x}} = A\mathbf{x} + B\mathbf{u}$, $\mathscr{T}(t) = \mathbf{0}$ with A, B constant, $\mathbf{u}(t) \in \Omega$ and piecewise continuous, and

$$C[\mathbf{u}(\cdot)] = \int_0^{t_1} |\mathbf{u}(t)|\, dt, \qquad |\mathbf{u}(t)| = \sum_{j=1}^{m} |u^j(t)|.$$

Show that for an optimal control, $\mathbf{u}(\cdot)$, the expression

$$-\sum_{j=1}^{n} |v^j| + \mathbf{v}^T B^T \mathbf{w}(t) + \mathbf{w}T(t)A\mathbf{x}.$$

is maximized when $\mathbf{v} = \mathbf{u}(t)$. Deduce from this that

$$u^j(t) = \operatorname{dez}\{[B^T\mathbf{w}]^j\}$$

where the *dead-zone sign function* is defined by

$$\operatorname{dez}\{\alpha\} = \begin{cases} \operatorname{sgn} \alpha, & \text{if } |\alpha| \ge 1, \\ 0, & \text{if } |\alpha| < 1. \end{cases}$$

7. (A Variation of Exercise 7.) We consider the same system as Exercise 7, except for a different cost function

$$C[\mathbf{u}(\cdot)] = \frac{1}{2}\int_0^{t_1} \mathbf{u}^T(t)\mathbf{u}(t)\, dt = \frac{1}{2}\sum_{j=1}^{m} \int_0^{t_1} |u^j(t)|^2\, dt.$$

Show that an optimal control $\mathbf{u}(\cdot)$ must make

$$-\tfrac{1}{2}\mathbf{v}^T\mathbf{v} + \mathbf{v}^T B^T \mathbf{w}(t)$$

a maximum when $\mathbf{v} = \mathbf{u}(t)$. Show that

$$u^j(t) = \operatorname{sat}\{[B^T\mathbf{w}(t)]^j\}, \qquad j = 1, \ldots, m$$

where the *saturation function* is defined by

$$\operatorname{sat}\{\alpha\} = \begin{cases} \alpha, & \text{if } |\alpha| < 1, \\ \operatorname{sgn} \alpha, & \text{if } |\alpha| \ge 1. \end{cases}$$

8. Carry through the complete synthesis of the optimal control for (a) Exercise 7, and (b) Exercise 8 when $(n = 1, m = 2)$

$$A = 1, \qquad B = [2, -1].$$

9. (*Much* harder than Exercise 9.) Carry through the complete synthesis of the optimal control for Exercise 7 when ($n=2$, $m=1$)

$$A=\begin{bmatrix}1 & -1\\ 0 & 1\end{bmatrix},\qquad B=\begin{bmatrix}-1\\ 1\end{bmatrix}.$$

Sketch the switching curves: responses which go through (0, 0) for 3 cases: $u=0$, $u=+1$, $u=-1$.

10. (The PMP is not sufficient.) Consider the scalar problem $\dot{x}=u$ with

$$C[u(\cdot)]=\int_0^{t_1}[u(t)]^2\{1-2x[t]u(t)+t[u(t)]^2\}\,dt,$$

$x_0=0$, $\mathscr{T}(t)\equiv 0$. We will use piecewise continuous controls with values $u(t)$ unrestricted. Obviously, one can take $u(t)\equiv 0$, $x[t]\equiv 0$, which gives a cost of zero. Show that this solution satisfies the PMP with $w^0(t)\equiv -1$, $w^1(t)\equiv 0$. Show that it is not optimal. (*Hint*: Let $h>0$ and

$$u_h(t)=\begin{cases}1/h, & 0\le t<h,\\ -1/(1-h), & h\le t<1.\end{cases}$$

Show that $\lim_{h\to 0+} C[u_h(\cdot)]=-\infty$). Carry through a similar example when $u(t)$ is restricted by $u(t)\in[-1,1]$. (*Hint*: $C[u_{h_0}(\cdot)]<0$, for some h_0, and $|u_{h_0}(t)|<K$ for some K. Now modify f^0 and/or f.)

11. (Continuing Exercise 2 from Chapter I.) A boat travels at velocity $\mathbf{v}(t)$, $|\mathbf{v}(t)|=1$, relative to a river whose current moves at velocity $\mathbf{c}$. Use the PMP to decide how to steer from a specified P_0 to a specified P_1 (see figure), in minimum time. (Let $|\mathbf{c}|=c$.) For a steering angle of γ relative to $\mathbf{c}$, derive

$$\dot{x}^1=c+|\mathbf{v}|\cos\gamma=c+\cos\gamma$$

$$\dot{x}^2=|\mathbf{v}|\sin\gamma=\sin\gamma.$$

Let $\mathbf{u}(t)=\left[\begin{smallmatrix}\cos\gamma\\ \sin\gamma\end{smallmatrix}\right]$. Describe Ψ, form the Hamiltonian H. Show that $\hat{\mathbf{w}}$ is constant. Find $M(\mathbf{x},\hat{\mathbf{w}})$ and show that M is attained for $\tan\gamma=\text{constant}$. Find the constant. (*Hint*: $M=0$).

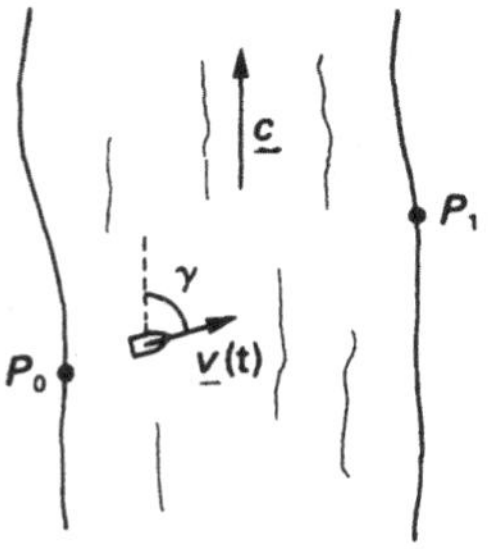

Figure E.12

12. (Continuing Exercise 12.) The situation is the same as in Exercise 12, except the target P_1 is replaced by a circle of radius r centered at (ξ^1, ξ^2):

$$\mathscr{T}(t) = \{(x^1, x^2) | [x^1 - \xi^1]^2 + [x^2 - \xi^2]^2 = r^2\}.$$

Show that the time-optimal trajectory must be normal to the target circle, thus the optimal response moves on the straight line from P_0 to (ξ^1, ξ^2).

13. (An example with t_1 specified, $\mathscr{T}(t) = R^n$.) The state of a servomechanism is described by

$$\dot{\mathbf{x}} = \mathbf{f}(\mathbf{x}) + B\mathbf{u}, \qquad \mathbf{x}[0] = \mathbf{x}_0, \qquad \mathbf{u}(t) \in \Omega$$

with $\mathbf{u}(\cdot)$ piecewise continuous, and t_1 *specified*, $\mathbf{x}[t_1]$ *not specified* $(\mathscr{T}(t) \equiv R^n.)$ Let the cost be measured by the *deviation from a given fixed desired state* $\mathbf{x}_{\#}[\cdot]$:

$$C[\mathbf{u}(\cdot)] = \int_0^{t_1} f_0(\mathbf{x}_{\#}[s] - \mathbf{x}[s])\, ds$$

with f^0 a given function from R^n into R. For example, we might be dealing with a heating system in a large space station (or greenhouse), with $\mathbf{u}(\cdot)$ a list of (time-varying) settings for various thermostats and $\mathbf{x}_{\#}[s]$ a given list of the desired temperatures at various locations. Then S_1 could be the fixed period of time over which the spaceship would be operating. Show that transversality implies $\mathbf{w}(t_1) = \mathbf{0}$. (*Hint*: $S_1 = R^n$.) Form the Hamiltonian, and show that an optimal control must be bang–bang or some combination of coasting with bang–bang. However, to obtain the state and costate vectors, show that we must solve a two-point boundary value problem involving a system of $2n+2$ differential equations. (A two-point boundary value problem is a system of ordinary differential equations on which some unknown functions are specified at $t = 0$, and the remaining unknowns are specified at $t = t_1$. This type of problem is intrinsically much more difficult to solve than an initial value problem.)

14. (Using change of variables in linear problems.) Consider the time-optimal control problem

$$\ddot{p} + 3\dot{p} + 2p = u(t)$$

with $p[0]$ and $\dot{p}[0]$ specified, $|u(t)| \leq 1$, $u(\cdot)$ piecewise continuous, and target $p[t_1] = \dot{p}[t_1] = 0$. Write this as a system $\dot{\mathbf{z}} = A\mathbf{z} + \mathbf{L}u$ in the obvious way. Show that if

$$Q = \begin{bmatrix} 1 & 1 \\ -1 & -2 \end{bmatrix},$$

then $D = Q^{-1}AQ$ is diagonal. Let $\mathbf{y}[\cdot] = Q^{-1}\mathbf{z}[\cdot]$ and derive the control problem for $\mathbf{y}[\cdot]$. Show that the resulting system is normal, and prove that there exists a time-optimal control. Derive the Hamiltonian and show that an optimal control must satisfy

$$u(t) = \operatorname{sgn}[w^1(t) + 2w^2(t)].$$

Solve $(\widehat{\text{Adj}}.)$. Solve the state equations for $u(t) \equiv +1$ and $u(t) \equiv -1$, and plot the two responses, Q_+ and Q_- (in **y**-space), which pass through **0**. Show that these are switching curves for optimal responses when $\mathbf{x}_0$ does not lie on $Q_+ \cup Q_-$.

15. (Change of Variable When A Cannot Be Diagonalized.) Consider the time-optimal control problem

$$\dddot{p} + \alpha^2 \ddot{p} = u \qquad (\alpha^2 > 0 \text{ a given constant}),$$

with $|u(t)| \leq 1$, $u(\cdot)$ piecewise continuous. Let $\mathbf{y} = [p, \dot{p}, \ddot{p}]^T$, and write the problem as a system, $\dot{\mathbf{y}} = A\mathbf{y} + \mathbf{b}u$. If

$$Q = \begin{bmatrix} 1 & 0 & \alpha^{-4} \\ 0 & 1 & -\alpha^{-2} \\ 0 & 0 & 1 \end{bmatrix},$$

show that the change of variable $\mathbf{z} = Q^{-1}\mathbf{y}$ gives a system $\dot{\mathbf{z}} = J\mathbf{z} + Q^{-1}\mathbf{b}u$ with J in Jordan canonical form. If we define

$$\mathbf{x} = \begin{bmatrix} \alpha^6 & 0 & 0 \\ 0 & \alpha^4 & 0 \\ 0 & 0 & \alpha^2 \end{bmatrix} \mathbf{z}$$

then derive the problem for $\mathbf{x}[\cdot]$. Form the Hamiltonian and show that an optimal control must satisfy

$$u(t) = -\operatorname{sgn}[w^1(t) - w^2(t) - w^3(t)].$$

Solve ($\widehat{\text{Adj}}$.) and argue that $u(\cdot)$ switches at most twice.

16. (The Rocket Car with a New Target.)
    (a) Use the PMP (with the transversality conditions) to solve the problem of bringing the rocket car to rest, i.e., $\mathscr{T}(t) = \{(x^1, 0) | x^1 \in R\}$.
    (b) Same as (a) except $\mathscr{T}(t) = \{(x^1, 0) | |x^1| \leq d\}$, d given.

17. (Singular Control Problem.) We say that an optimal control (or the problem from which it stems) is *singular* if maximizing the Hamiltonian does not explicitly determine $\mathbf{u}(t)$ as a single-valued function of $\mathbf{x}[t]$ and $\mathbf{w}(t)$. Consider the optimal control problem

$$\mathbf{x} = \begin{bmatrix} p \\ q \end{bmatrix}, \qquad \dot{p} = q, \qquad \dot{q} = -\tfrac{1}{3}[q]^3 + u,$$

$$C[u(\cdot)] = \int_0^{t_1} [\lambda_1 + |u(t)|]\, dt, \qquad \mathscr{T}(t) \equiv \mathbf{0},$$

with λ_1 given, $0 < \lambda_1 < 2$, $|u(t)| \leq 1$ and piecewise continuous. Form the Hamiltonian and the system ($\widehat{\text{Adj}}$.). Show that

$$|w^2(t)| < 1 \Rightarrow u(t) = 0,$$
$$|w^2(t)| > 1 \Rightarrow u(t) = \operatorname{sgn} w^2(t).$$

$$w^2(t) = +1 \Rightarrow 0 \leq u(t) \leq +1,$$
$$w^2(t) = -1 \Rightarrow -1 \leq u(t) \leq 0.$$

Thus the problem is singular, since $u(t)$ is not specified when $|w^2(t)| = 1$. Supposing that $|w^2(t)| = 1$ on some interval I, show that $q[t] \equiv$ constant and $u(t) = [q]^3/3$ on I.

18. (How to Handle a Problem with a Fixed Time Interval.) Consider the scalar problem $\dot{p} = -u$ for $0 \leq t \leq 1$, with $p[0] = 0$, $p[1] = 0$ both specified,

$$C[u(\cdot)] = \int_0^1 (u+1)\,dt,$$

and $u(t) \in \Omega$ with $u(\cdot)$ piecewise continuous.

(a) Show that any control for which $\int_0^1 u(t)\,dt = 0$ is optimal. Attempt to apply the PMP directly and derive $w(t) \equiv 0$.

(b) To use the PMP on the above problem, introduce the new state variable $q[\cdot]$ with

$$\dot{q} = 1, \qquad q(0) = 0.$$

Then define

$$\mathbf{x} = \begin{bmatrix} p \\ q \end{bmatrix}, \quad \text{with } \mathbf{x}[0] = \mathbf{0}, \quad \mathbf{x}[t_1] = \begin{bmatrix} 0 \\ 1 \end{bmatrix}.$$

Show that the new problem for $\mathbf{x}[\cdot]$ is equivalent to the original problem for $p[\cdot]$. Use the PMP to synthesize the optimal control.

Appendix to Chapter V: A Proof of the Pontryagin Maximum Principle

We consider the problem

$$\dot{\mathbf{x}} = \mathbf{f}(\mathbf{x}, \mathbf{u}), \qquad \mathbf{x}[0] = \mathbf{x}_0, \qquad \mathbf{u}(\cdot) \text{ measurable,}$$

with $\mathbf{x}[t] \in R^n$, $\mathbf{u}(t) \in \Psi \subset R^m$ with Ψ an arbitrary bounded set. We assume that $\mathbf{f}$ is continuous in $(\mathbf{x}, \mathbf{u})$ and continuously differentiable in $\mathbf{x}$. Finally we assume a fixed target $\mathscr{T}(t) \equiv \mathbf{x}_1$ and a cost function

$$C[\mathbf{u}(\cdot)] = \int_0^{t_1} f^0(\mathbf{x}[s], \mathbf{u}(s))\, ds$$

with f^0 bounded below on $R^n \times \Psi$, continuous in $(\mathbf{x}, \mathbf{u})$, and continuously differentiable in $\mathbf{x}$. Here t_1 is the (unspecified) time of arrival at the target.

We define a regular point of a control $\mathbf{u}(\cdot)$ as a point t at which

$$\lim_{h \to 0} \frac{1}{h} \int_{t-h}^{t} [\hat{\mathbf{f}}(\mathbf{x}[\tau], \mathbf{u}(\tau)) - \hat{\mathbf{f}}(\mathbf{x}[t], \mathbf{u}(t))]\, d\tau = 0.$$

Under our assumptions, almost all points are regular.

The reader who prefers to avoid measure theory can assume piecewise constant controls, and use non-jump points as regular points. In either case, the set of non-regular points has measure zero, i.e., almost all points are regular.

Given a particular control-response pair $(\hat{\mathbf{x}}[t], \mathbf{u}(t))$ we can form $\hat{\mathbf{f}}_{\hat{\mathbf{x}}}[t] \equiv [\partial f^i/\partial x^j]$ evaluated along $(\hat{\mathbf{x}}[t], \mathbf{u}(t))$, $i = 0, 1, 2, \ldots, n$; $j = 0, 1, \ldots, n$. We can then form the associated linear and adjoint systems:

$(\widehat{\text{Lin.}})$ $$\dot{\hat{\mathbf{y}}} = \hat{\mathbf{f}}_{\hat{\mathbf{x}}}[t]\hat{\mathbf{y}},$$

$(\widehat{\text{Adj.}})$ $$\dot{\hat{\mathbf{w}}} = -\{\hat{\mathbf{f}}_{\hat{\mathbf{x}}}[t]\}^T \hat{\mathbf{w}}.$$

We can then define the Hamiltonian associated with any control-response pair and adjoint solution $\hat{\mathbf{w}}(\cdot)$:

$$\mathcal{H}(\hat{\mathbf{w}}, \hat{\mathbf{x}}, \mathbf{u}) = \langle \hat{\mathbf{w}}, \hat{\mathbf{f}}(\mathbf{x}, \mathbf{u})\rangle = \sum_{i=0}^{n} w^i(t) f^i(\mathbf{x}[t], \mathbf{u}(t)),$$

and its maximal function

$$\mathcal{M}(\hat{\mathbf{w}}, \hat{\mathbf{x}}) = \sup_{\mathbf{v}\in\Psi} \mathcal{H}(\hat{\mathbf{w}}, \hat{\mathbf{x}}, \mathbf{v}).$$

Theorem. *If* $(\hat{\mathbf{x}}_*[\cdot], \mathbf{u}_*(\cdot))$ *is an optimal pair for the control problem described above, then there exists a solution* $\hat{\mathbf{w}}(\cdot)$ *of* $(\widehat{Adj.})$ *such that almost everywhere*

$$\mathcal{H}(\hat{\mathbf{w}}(t), \hat{\mathbf{x}}[t], \mathbf{u}_*(t)) \equiv \mathcal{M}(\hat{\mathbf{w}}(t), \hat{\mathbf{x}}_*[t]).$$

Furthermore, $\mathcal{M}(\hat{\mathbf{w}}(t), \hat{\mathbf{x}}_*[t]) \equiv 0$ *on* $[0, t_1]$.

Remarks

(1) The extension to non-autonomous problems is relatively straightforward, if one introduces an additional state variable $x^{n+1}[\cdot]$, defined by $\dot{x}_{n+1} = 0$, $x_{n+1}[0] = 0$. This imbeds t in the state vector, and the new problem is autonomous. This approach requires $\hat{\mathbf{f}}(t, \mathbf{x}, \mathbf{u})$ to be continuously differentiable in t in order to make the new system continuously differentiable in x^{n+1}. One can prove the PMP under weaker assumptions by a direct attack on the non-autonomous problem – see Lee and Markus [1967], Chapter 5, or Berkovitz [1974], Chapter VI.

(2) The statement and proof of the PMP is more complicated when the target $\mathcal{T}(t)$ is a closed time-varying set. We will not deal with this case. The interested reader can find the details in Lee and Markus or Berkovitz (op cit.).

Proof

Our proof of the PMP will occupy most of the remainder of this appendix. At the end of this appendix, we describe other approaches to the proof. Our approach is essentially that of the original proof of Pontryagin *et al.* [1964]; the presentation in Lee and Markus [1967] is based on this same approach. We follow in part the presentation in a paper of Diliberto which appears in Leitmann [1977], and the presentation in Lee and Markus.

The key idea is to perturb an optimal control $\mathbf{u}_*(\cdot)$ by changing its value to any admissible vector $\mathbf{v}$ over any small time interval (Figure 1):

$$\mathbf{u}_\varepsilon(t) = \begin{cases} \mathbf{u}_*(t), & t \notin [\tau_i - \varepsilon c^i, \tau_i - \varepsilon k^i], \\ \mathbf{v}, & t \in [\tau_i - \varepsilon c^i, \tau_i - \varepsilon k^i], \quad \mathbf{v} \in \Psi. \end{cases}$$

Here τ_i is any instant in (t_0, t_1), and c^i, k^i, are any non-negative constants. In Lemma 1 we show that if $\hat{\mathbf{x}}_\varepsilon[\cdot]$ is the response associated with $\mathbf{u}_\varepsilon(\cdot)$, then

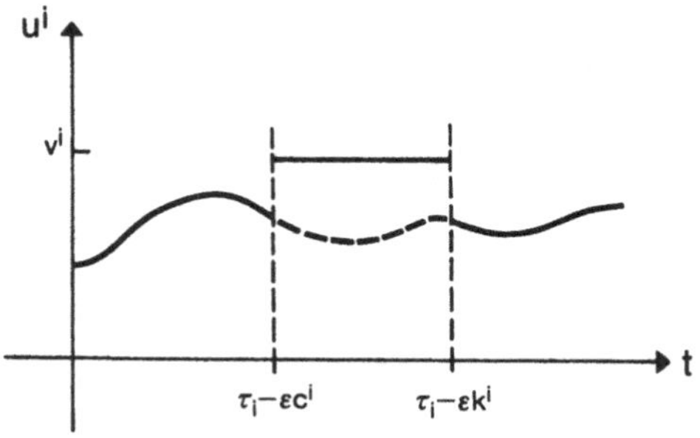

Figure 1 The ith Components of $\mathbf{u}_*(\cdot)$ and $\mathbf{u}_\varepsilon(\cdot)$

$$\hat{\mathbf{x}}_\varepsilon[\tau_i]=\hat{\mathbf{x}}_*[\tau_i]+\varepsilon c^i\hat{\boldsymbol{\zeta}}+o(\varepsilon),$$
$$\hat{\boldsymbol{\zeta}}=\hat{\mathbf{f}}(\mathbf{x}_*[\tau_i],\mathbf{v})-\hat{\mathbf{f}}(\mathbf{x}_*[\tau_i],\mathbf{u}_*(\tau_i)).$$

In Lemma 2 we show that this perturbation in the response at τ_i propagates to time t_1 under ($\widehat{\text{Lin.}}$). (The fact that $\hat{\mathbf{x}}_\varepsilon[\cdot]$ exists on $[0, t_1]$ for ε small follows from standard results):

$$\hat{\mathbf{x}}_\varepsilon[t_1]=\hat{\mathbf{x}}_*[t_1]+\varepsilon c^i Y(t_1,\tau_i)\hat{\boldsymbol{\zeta}}+o(\varepsilon),$$

where $Y(t,\tau_i)$ is the fundamental matrix for ($\widehat{\text{Lin.}}$) satisfying $Y(\tau_i,\tau_i)=I$ (thus $\hat{\mathbf{z}}[t]=Y(t,\tau_i)\hat{\boldsymbol{\zeta}}$ solves ($\widehat{\text{Lin.}}$) with $\hat{\mathbf{z}}[\tau_i]=\hat{\boldsymbol{\zeta}}$).

If we select a distinct set of times τ_i: $t_0<\tau_1<\tau_2<\cdots<\tau_p<t_1$, and perturb $\mathbf{u}_*(\cdot)$ near each τ_i by $\mathbf{v}_i$ as described above, then for any $t>\tau_p$ the resulting response can be written

$$\hat{\mathbf{x}}_\varepsilon[t]=\hat{\mathbf{x}}_*[t]+\varepsilon\sum c^i Y(t,\tau_i)\hat{\boldsymbol{\zeta}}_i+o(\varepsilon),$$

where $\hat{\boldsymbol{\zeta}}_i=\hat{\mathbf{f}}(\mathbf{x}_*[\tau_i],\mathbf{v}_i)-\hat{\mathbf{f}}(x_*[\tau_i],\mathbf{u}_*(\tau_i))$. Thus the effect at time t_1 of all possible such perturbations is, to first order in ε, a cone (recall $c^i\geq 0$) formed from the *elementary state perturbation vectors*

$$\hat{\mathbf{z}}_i=Y(t_1,\tau_i)\hat{\boldsymbol{\zeta}}_i,\quad\text{where}\quad \hat{\boldsymbol{\zeta}}_i=\hat{\mathbf{f}}(\mathbf{x}_*[\tau_i],\mathbf{v}_i)-\hat{\mathbf{f}}(\mathbf{x}_*[\tau_i],\mathbf{u}_*(\tau_i)).$$

This leads to the following definitions:

(i) $$k(\tau)=\left\{\sum_{i=1}^{p}c^i\hat{\boldsymbol{\zeta}}_i \,|\, c^i\geq 0,\ \hat{\boldsymbol{\zeta}}_i=\hat{\mathbf{f}}(\mathbf{x}_*[\tau],\mathbf{v}_i)-\hat{\mathbf{f}}(\mathbf{x}_*[\tau],\mathbf{u}_*(\tau)),\ \mathbf{v}_i\in\Psi,\ p\text{ any natural number}\right\},$$

(ii) $$\mathscr{K}(t)=\left\{\sum_{i=1}^{p}c^iY(t,\tau_i)\hat{\boldsymbol{\zeta}}_i \,|\, c^i\geq 0,\ \tau_i\in(0,t_1),\ \hat{\boldsymbol{\zeta}}_i\in k(\tau_i)\right\}.$$

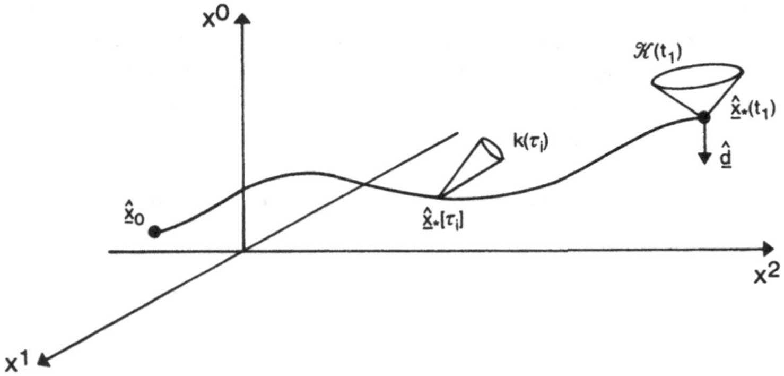

Figure 2

At a regular point t, the cone $\mathcal{K}(t)$ is only an approximation to the reachable set at time t, but it is adequate for our purposes. If $(\mathbf{u}_*(\cdot), \mathbf{x}_*[\cdot])$ is optimal, then $\mathcal{K}(t)$ is either of dimension less than $n+1$ and/or it does not contain any vertical downward vector $\hat{\mathbf{d}} = \mu(1, 0, \ldots, 0)^T$, $\mu < 0$, at $\mathbf{x}_*[t_1]$. This is the content of Lemma 3. This lemma is the most difficult part of the proof of the PMP, in that it requires a variant of Brouwer Fixed Point Theorem.

As a consequence of Lemma 3, if t_1 is regular we can separate R^{n+1} into two halfspaces by means of a support hyperplane at the vertex of $\mathcal{K}(t_1)$. We choose a vector $\hat{\mathbf{w}}_1$ normal to this plane, pointing away from the cone $\mathcal{K}(t_1)$. Then $\hat{\mathbf{w}}_1 \cdot \hat{\mathbf{z}} \leq 0$ for all $\hat{\mathbf{z}} \in \mathcal{K}(t_1)$. We then define a solution of $(\widehat{\text{Adj}}.)$ by $\hat{\mathbf{w}}(t_1) = \hat{\mathbf{w}}_1$. Then, almost everywhere

$$\hat{\mathbf{w}}[t] = \{[Y(t, t_1)]^T\}^{-1}\hat{\mathbf{w}}_1.$$

(If t_1 is not regular, we replace it in the above argument by any regular point from $(0, t_1)$.) Here as usual $Y(t, \tau)$ is the matrix solution (as a function of t) of $(\widehat{\text{Lin}}.)$ satisfying $Y(\tau, \tau) = I$. (Here we use the familiar fact that if $Y(t, \tau)$ is a fundamental matrix for $(\widehat{\text{Lin}}.)$, then $[Y^T]^{-1}$ is a fundamental matrix for $(\widehat{\text{Adj}}.)$.)

Assuming the validity of the lemmas described above, we can complete the proof of the first part of the PMP as follows. At any regular instant t,

$$\begin{aligned}\mathcal{H}(\hat{\mathbf{w}}(t), \hat{\mathbf{x}}_*[t], \mathbf{v}) - \mathcal{H}(\hat{\mathbf{w}}(t), \hat{\mathbf{x}}_*[t], \mathbf{u}_*(t)) &= \langle \hat{\mathbf{w}}(t), \hat{\mathbf{f}}(\mathbf{x}_*[t], \mathbf{v}) - \hat{\mathbf{f}}(\mathbf{x}_*[t], \mathbf{u}_*(t))\rangle \\ &= \langle \{[Y(t, t_1)]^T\}^{-1}\hat{\mathbf{w}}_1, \hat{\boldsymbol{\zeta}}\rangle \\ &= \langle \hat{\mathbf{w}}_1, \hat{\mathbf{z}}\rangle \quad \text{a.e.,}\end{aligned}$$

where $\hat{\boldsymbol{\zeta}} = [\hat{\mathbf{f}}(\mathbf{x}_*[t], \mathbf{v}) - \hat{\mathbf{f}}(\mathbf{x}_*[t], \mathbf{u}_*(t))] \in k(t)$, and $\hat{\mathbf{z}} = [Y(t, t_1)]^{-1}\hat{\boldsymbol{\zeta}} \in \mathcal{K}(t_1)$. Therefore, $\langle \hat{\mathbf{w}}_1, \hat{\mathbf{z}}\rangle \leq 0$, i.e.,

$$\mathcal{H}(\hat{\mathbf{w}}(t), \hat{\mathbf{x}}_*[t], \mathbf{v}) \leq \mathcal{H}(\hat{\mathbf{w}}(t), \hat{\mathbf{x}}_*[t], \mathbf{u}_*(t)).$$

Since $\mathbf{v} \in \Psi$ was arbitrary, this shows that

$$\mathcal{H}(\hat{\mathbf{w}}(t), \hat{\mathbf{x}}_*[t], \mathbf{u}_*(t)) = \mathcal{M}(\hat{\mathbf{w}}(t), \mathbf{x}_*[t]), \quad \text{a.e.}$$

To see that $\mathcal{M}$ is in fact zero requires additional efforts, and these are described in detail in Lemmas 4 and 5, and the subsequent discussion.

The reader should note that the PMP is valid for any *extremal* solution, that is, any solution which lies on the boundary of the attainable set at time t_1 in the extended state space.

Lemma 1. *For positive constants k, c, and ε, let* $\mathbf{u}_\varepsilon(\cdot)$ *be defined by*

$$\mathbf{u}_\varepsilon(t)=\begin{cases}\mathbf{u}_*(t), & t\in[a,\sigma_1)\cup[\sigma_2,\tau],\\ \mathbf{v}, & t\in[\sigma_1,\sigma_2),\end{cases}$$

where $\sigma_1=\tau-\varepsilon(c+k)$, $\sigma_2=\tau-\varepsilon k$, $\mathbf{v}\in\Psi$.

If $\mathbf{x}_\varepsilon[\cdot]$ *is the corresponding response, then at all regular points*

(1) $\hat{\mathbf{x}}_*[t]=\hat{\mathbf{x}}_*[\sigma_1]+(t-\sigma_1)\hat{\mathbf{f}}(\mathbf{x}_*[\tau],\mathbf{u}_*(\tau))+o(\varepsilon)$, *on* $[\sigma_1,\tau]$;

(2) $\hat{\mathbf{x}}_\varepsilon[t]=\hat{\mathbf{x}}_\varepsilon[\sigma_1]+(t-\sigma_1)\hat{\mathbf{f}}(\mathbf{x}_\varepsilon[\tau],\mathbf{v})+o(\varepsilon)$, *on* $[\sigma_1,\sigma_2]$;

(3) $\hat{\mathbf{x}}_\varepsilon[t]=\hat{\mathbf{x}}_\varepsilon[\sigma_1]+\varepsilon c\hat{\mathbf{f}}(\mathbf{x}_\varepsilon[\tau],\mathbf{v})+(t-\sigma_2)\hat{\mathbf{f}}(\mathbf{x}_*[\tau],\mathbf{u}_*(\tau))+o(\varepsilon)$ *on* $[\sigma_2,\tau]$;

(4) $\hat{\mathbf{x}}_\varepsilon[\sigma_1]=\hat{\mathbf{x}}_*[\sigma_1]$;

(5) $|\hat{\mathbf{x}}_\varepsilon[t]-\hat{\mathbf{x}}_*[t]|\le 2(t-\sigma_1)M$ *for* $t\in[\sigma_1,\tau]$;

(6) $\hat{\mathbf{x}}_\varepsilon[\tau]-\hat{\mathbf{x}}_*[\tau]=\varepsilon c[\hat{\mathbf{f}}(\hat{\mathbf{x}}_*[\tau],\mathbf{v})-\hat{\mathbf{f}}(\hat{\mathbf{x}}_*[\tau],\mathbf{u}_*(\tau))]+o(\varepsilon)$.

Remarks

(1) Equation (6) shows that the perturbation at time τ is differentiable in ε, even though $\hat{\mathbf{f}}(\mathbf{x},\mathbf{u})$ is only continuous in ε.
(2) Equation (1) holds for any response control pair; optimality does not play a role here.

Proof. (4) is obvious.

Since $|\hat{\mathbf{f}}|\le K^*$ on $R^n\times\Psi$, for any response we have

$$|\hat{\mathbf{x}}[\tau_2]-\hat{\mathbf{x}}[\tau_1]|=\left|\int_{\tau_1}^{\tau_2}\hat{\mathbf{f}}\right|\le K^*(\tau_2-\tau_1).$$

Therefore,

(i) $|\hat{\mathbf{x}}_*[t]-\hat{\mathbf{x}}_*[\sigma_1]|\le K^*(t-\sigma_1)$,
(ii) $|\hat{\mathbf{x}}_\varepsilon[t]-\hat{\mathbf{x}}_\varepsilon[\sigma_1]|\le K^*(t-\sigma_1)$,

so

$$|\hat{\mathbf{x}}_\varepsilon[t]-\hat{\mathbf{x}}_*[t]|\le|\hat{\mathbf{x}}_\varepsilon[t]-\hat{\mathbf{x}}_\varepsilon[\sigma_1]|+|\hat{\mathbf{x}}_*[\sigma_1]-\hat{\mathbf{x}}_*[t]|\le 2K^*(t-\sigma_1)$$

and (5) holds.

To establish (1), we write

$$\begin{aligned}
\hat{\mathbf{x}}_*[t] &= \hat{\mathbf{x}}_*[\sigma_1] + \int_{\sigma_1}^{t} \hat{\mathbf{f}}(\mathbf{x}_*[s], \mathbf{u}_*(s))\, ds \\
&= \hat{\mathbf{x}}_*[\sigma_1] + \int_{\sigma_1}^{t} \hat{\mathbf{f}}(\mathbf{x}_*[\tau], \mathbf{u}_*(\tau))\, ds \\
&\quad + \int_{\sigma_1}^{t} [\hat{\mathbf{f}}(\mathbf{x}_*[s], \mathbf{u}_*(s)) - \hat{\mathbf{f}}(\mathbf{x}_*[\tau], \mathbf{u}_*(s))]\, ds \\
&\quad + \int_{\sigma_1}^{t} [\hat{\mathbf{f}}(\mathbf{x}_*[\tau], \mathbf{u}_*(s)) - \hat{\mathbf{f}}(\mathbf{x}_*[\tau], \mathbf{u}_*(\tau))]\, ds \\
&= \hat{\mathbf{x}}_*[\sigma_1] + \hat{\mathbf{f}}(\mathbf{x}_*[\tau], \mathbf{u}_*(\tau))(t-\sigma_1) + \mathcal{I}_1 + \mathcal{I}_2.
\end{aligned}$$

Since $\hat{\mathbf{f}}$ is continuously differentiable with respect to $\mathbf{x}$, it is locally Lipschitzian with respect to $\mathbf{x}$, so

$$\begin{aligned}
|\mathcal{I}_1| &\le \int_{\sigma_1}^{t} K|\mathbf{x}_*[s] - \mathbf{x}_*[\tau]|\, ds \le \int_{\sigma_1}^{t} K\left|\int_{s}^{\tau} \hat{\mathbf{f}}(\mathbf{x}_*[r], \mathbf{u}_*(r))\, dr\right| ds \\
&\le \int_{\sigma_1}^{t} KK^*(\tau - s)\, ds \le KK^*(\tau-\sigma_1)^2 = o(\varepsilon).
\end{aligned}$$

At a regular point τ,

$$\lim_{\sigma_1\to\tau} \frac{1}{\tau-\sigma_1}\int_{\sigma_1}^{\tau} |\hat{\mathbf{f}}(\mathbf{x}_*[\tau], \mathbf{u}_*(s)) - \hat{\mathbf{f}}(\mathbf{x}_*[\tau], \mathbf{u}_*(\tau))|\, ds = 0,$$

which implies that $\mathcal{I}_2 = o(1)(\tau-\sigma_1) = o(\varepsilon)$. This proves (1).

To establish (2) and (3), we write

$$\begin{aligned}
\hat{\mathbf{x}}_\varepsilon[t] &= \hat{\mathbf{x}}_\varepsilon[\sigma_1] + \int_{\sigma_1}^{t} \hat{\mathbf{f}}(\mathbf{x}_\varepsilon[s], \mathbf{u}(s))\, ds \\
&= \hat{\mathbf{x}}_\varepsilon[\sigma_1] + \int_{\sigma_1}^{t} \hat{\mathbf{f}}(\mathbf{x}_\varepsilon[\tau], \mathbf{v})\, ds + \int_{\sigma_1}^{t} [\hat{\mathbf{f}}(\mathbf{x}_\varepsilon[s], \mathbf{u}_\varepsilon(s)) \\
&\quad - \hat{\mathbf{f}}(\mathbf{x}_\varepsilon[\tau], \mathbf{u}_\varepsilon(s))]\, ds + \int_{\sigma_1}^{t} [\hat{\mathbf{f}}(\mathbf{x}_\varepsilon[\tau], \mathbf{u}_\varepsilon(s)) - \hat{\mathbf{f}}(\mathbf{x}_\varepsilon[\tau], \mathbf{v})]\, ds
\end{aligned}$$

for $\sigma_1 \le t \le \tau$.

The last two integrals on the right side are $o(\varepsilon)$ by arguments exactly as above, so

$$\hat{\mathbf{x}}_\varepsilon[t] = \hat{\mathbf{x}}[\sigma_1] + (t-\sigma_1)\hat{\mathbf{f}}(\mathbf{x}_\varepsilon[\tau], \mathbf{v}) + o(\varepsilon), \qquad \sigma_1 \le t \le \tau.$$

In particular,

$$\hat{\mathbf{x}}_\varepsilon[\sigma_2] = \hat{\mathbf{x}}_\varepsilon[\sigma_1] + \varepsilon c\hat{\mathbf{f}}(\mathbf{x}_\varepsilon[\tau], \mathbf{v}) + o(\varepsilon). \tag{7}$$

If we now begin with $\hat{\mathbf{x}}_\varepsilon[t]=\hat{\mathbf{x}}_\varepsilon[\sigma_2]+\int_{\sigma_2}^{t}\hat{\mathbf{f}}(\mathbf{x}_\varepsilon[s],\mathbf{u}_\varepsilon(s))\,ds$, and parallel the derivation of (1) and (2), we get

$$\hat{\mathbf{x}}_\varepsilon[t]=\hat{\mathbf{x}}_\varepsilon[\sigma_2]+(t-\sigma_2)\hat{\mathbf{f}}(\mathbf{x}_\varepsilon[\tau],\mathbf{u}(\tau))+o(\varepsilon),$$

and this combined with (7) gives (3).

Finally, we obtain (6) as follows. From (1) and (3) we have: (recall $\hat{\mathbf{x}}_*[\sigma_1]=\hat{\mathbf{x}}_\varepsilon[\sigma_1]$)

$$\hat{\mathbf{x}}_*[\tau]=\hat{\mathbf{x}}_*[\sigma_1]+\varepsilon(c+k)\hat{\mathbf{f}}(\mathbf{x}_*[\tau],\mathbf{u}_*(\tau))+o(\varepsilon)$$

$$\hat{\mathbf{x}}_\varepsilon[\tau]=\hat{\mathbf{x}}_\varepsilon[\sigma_1]+\varepsilon c\hat{\mathbf{f}}(\mathbf{x}_\varepsilon[\tau],\mathbf{v})+\varepsilon k\hat{\mathbf{f}}(\mathbf{x}_\varepsilon[\tau],\mathbf{u}_*(\tau))+o(\varepsilon).$$

But from (5), we have $\hat{\mathbf{x}}_\varepsilon[\tau]=\hat{\mathbf{x}}_*[\tau]+o(\varepsilon)$, so we can replace $\hat{\mathbf{x}}_\varepsilon[\tau]$ by $\hat{\mathbf{x}}_*[\tau]$ on the right side of the last equation above to get

$$\hat{\mathbf{x}}_\varepsilon[\tau]=\hat{\mathbf{x}}_*[\sigma_1]+\varepsilon c\hat{\mathbf{f}}(\mathbf{x}_*[\tau],\mathbf{v})+\varepsilon k\hat{\mathbf{f}}(\mathbf{x}_*[\tau],\mathbf{u}_*(\tau))+o(\varepsilon).$$

Subtracting this from the first equation above, we obtain (6). □

Lemma 2. *Let* $\mathbf{u}(\cdot)$ *be any fixed admissible control, with response* $\hat{\mathbf{x}}[\cdot]$ *on* $[t_0, t_1]$. *Suppose that* $\hat{\mathbf{x}}_\varepsilon[\cdot]$ *solves the same differential equation as* $\hat{\mathbf{x}}[\cdot]$:

$$\dot{\hat{\mathbf{x}}}_\varepsilon=\hat{\mathbf{f}}(\mathbf{x}_\varepsilon,\mathbf{u}(t)),\quad \textit{on } [\tau, t_1]$$

with $\hat{\mathbf{x}}_\varepsilon[\tau]=\hat{\mathbf{x}}[\tau]+\varepsilon\hat{\boldsymbol{\zeta}}+o(\varepsilon)$, $\hat{\boldsymbol{\zeta}}$ *a fixed vector in* R^{n+1}. *Then*

$$\hat{\mathbf{x}}_\varepsilon[t]=\hat{\mathbf{x}}[t]+\varepsilon Y(t,\tau)\hat{\boldsymbol{\zeta}}+o(\varepsilon)$$

where $Y(t,\tau)$ *is the fundamental matrix for* $(\widehat{Lin.})$ *which satisfies* $Y(\tau,\tau)=I$.

Proof. This is a well-known folk theorem. The lemma asserts that for any $\hat{\boldsymbol{\zeta}}$, the function $\hat{\mathbf{x}}_\varepsilon[t]$ is differentiable with respect to ε and at $\varepsilon=0$ this derivative is

$$\frac{\partial}{\partial\varepsilon}(\hat{\mathbf{x}}_\varepsilon[t])|_{\varepsilon=0}=Y(t,\tau)\hat{\boldsymbol{\zeta}}.$$

To see this, we use the standard result (Coddington and Levinson [1955], Chapter 2) that the solution $\hat{\mathbf{y}}(t;\tau,\hat{\boldsymbol{\zeta}})$ of the initial-value problem

$$\dot{\hat{\mathbf{y}}}=\hat{\mathbf{f}}(\mathbf{y},\mathbf{u}(t)),\quad \hat{\mathbf{y}}(\tau)=\hat{\boldsymbol{\zeta}},\qquad \tau\le t\le t_1,$$

is differentiable with respect to its specified initial value $\hat{\mathbf{y}}(\tau)$, and this derivative is

$$\frac{\partial\hat{\mathbf{y}}}{\partial\hat{\mathbf{y}}(\tau)}=Y(t,\tau).$$

Then, by the chain rule,

$$\left.\frac{\partial\hat{\mathbf{y}}}{\partial\varepsilon}\right|_{\varepsilon=0}=\frac{\partial\hat{\mathbf{y}}}{\partial\hat{\mathbf{y}}(\tau)}\cdot\left.\frac{\partial\hat{\mathbf{y}}(\tau)}{\partial\varepsilon}\right|_{\varepsilon=0}=Y(t,\tau)\hat{\boldsymbol{\zeta}}.$$ □

Lemma 3. *If t is a regular point for the control-response pair* $(\hat{\mathbf{x}}[\cdot], \mathbf{u}(\cdot))$, *if the associated perturbation cone* $\mathcal{K}(t)$ *is such that* $\dim \mathcal{K}(t) = n+1$ *and the interior of* $\mathcal{K}(t)$ *contains a vertical downward vector* $\hat{\mathbf{d}} = \mu(1, 0, \ldots, 0)^T$, $\mu < 0$, *then* $(\hat{\mathbf{x}}[\cdot], \mathbf{u}(\cdot))$ *is not optimal.*

Proof. In Part A we show that if $(\hat{\mathbf{x}}[\cdot], \mathbf{u}(\cdot))$, $(\hat{\mathbf{y}}[\cdot], \mathbf{v}(\cdot))$ are control-response pairs and if for some instants τ, τ':

$$\mathbf{x}[\tau] = \mathbf{y}[\tau'], \qquad x^0[\tau] > y^0[\tau']$$

then $(\hat{\mathbf{x}}[\cdot], \mathbf{u}(\cdot))$ cannot be optimal. In Part B we show that under our assumptions such a $\hat{\mathbf{y}}[\cdot]$ exists.

Part A. Given the situation described above, with $\mathbf{x}[t_1] = \mathbf{x}_1$, we define

$$\mathbf{u}_{\#}(t) = \begin{cases} \mathbf{v}(t), & 0 \le t \le \tau'; \\ \mathbf{u}(t + \tau - \tau'), & \tau' < t \le t_1 + \tau' - \tau. \end{cases}$$

Clearly, the response $\hat{\mathbf{x}}_{\#}[t] \equiv \hat{\mathbf{y}}[t]$ for $0 \le t \le \tau'$. In particular, $\hat{\mathbf{x}}_{\#}[\tau'] = \hat{\mathbf{y}}[\tau'] = (y^0(\tau'), \mathbf{x}(\tau')^T)^T$.

We claim that $\hat{\mathbf{x}}_{\#}[t] = \hat{\mathbf{x}}[t + \tau - \tau'] + (c^0, 0, \ldots, 0)^T$ on $[\tau', t_1 + \tau' - \tau]$, where $c^0 = y^0[\tau'] - x^0[\tau] < 0$. This means that $\hat{\mathbf{x}}_{\#}[\cdot]$ is successful at lower cost:

$$\mathbf{x}_{\#}[t_1 + \tau' - \tau] = \mathbf{x}[t_1] = \mathbf{x}_1, \qquad x^0[t_1 + \tau' - \tau] = x^0[t_1] + c^0.$$

Recall first of all that $\hat{\mathbf{f}}$ does not depend on x^0. Therefore the function $\hat{\mathbf{x}}[t + \tau - \tau'] + (c^0, 0, \ldots, 0)^T$ solves $\dot{\hat{\mathbf{x}}} = \hat{\mathbf{f}}(\mathbf{x}, \mathbf{u})$. At $t = \tau'$,

$$\hat{\mathbf{x}}[\tau' + \tau - \tau'] + (c^0, 0, \ldots, 0)^T = (y^0(\tau'), \mathbf{x}(\tau)^T)^T.$$

This is exactly the value specified for $\hat{\mathbf{x}}_{\#}[\tau']$, and since solutions of initial-value problems are unique,

$$\hat{\mathbf{x}}_{\#}[t] \equiv \hat{\mathbf{x}}[t + \tau - \tau'].$$

Part B. Let $(\hat{\mathbf{x}}[\cdot], \mathbf{u}(\cdot))$ be a control-response pair. According to (6), the control change that replaces $\mathbf{u}(t)$ by the admissible constant vector $\mathbf{v}_i$ on $[\tau_i - \varepsilon(c^i + k^i), \tau_i - \varepsilon k^i]$ changes $\hat{\mathbf{x}}[\tau_i]$ by $\hat{\boldsymbol{\zeta}}_i = \varepsilon c^i[\hat{\mathbf{f}}(\mathbf{x}[\tau_i], \mathbf{v}_i) - \hat{f}(\mathbf{x}[\tau_i], u(\tau_i))]$, (to first order in ε). By Lemma 2, this change at τ_i propagates to a later time τ as $\hat{\mathbf{z}}_i = Y(\tau, \tau_i)\hat{\zeta}_i$. If we make a finite set of such perturbations at times $0 < \tau_1 < \tau_2 < \cdots < \tau_p < \tau$, then for sufficiently small $\varepsilon_0 > 0$, the corresponding response satisfies

$$\hat{\mathbf{x}}_\varepsilon[\tau, \mathbf{c}] = \hat{\mathbf{x}}[\tau] + \varepsilon \sum_{i=1}^{p} c^i \hat{\mathbf{z}}_i + o(\varepsilon), \qquad 0 < \varepsilon < \varepsilon_0.$$

Now if $\dim \mathcal{K}(\tau) = n+1$ and $\hat{\mathbf{d}} \in \mathcal{K}(\tau)$, then

$$\hat{\mathbf{d}} = \sum_{i=1}^{n+1} \alpha^i \hat{\mathbf{z}}_i, \qquad \alpha^i > 0,$$

for some independent set of vectors $\hat{\mathbf{z}}_i \in \mathcal{K}(\tau)$. Since $\hat{\mathbf{d}}$ is in the *interior* of $\mathcal{K}(\tau)$, all of the α^i's are positive. We want to show that there is an actual *response* $\hat{\mathbf{y}}[\cdot]$ such that $\hat{\mathbf{y}}[\tau] = \hat{\mathbf{x}}[\tau] + \hat{\mathbf{d}}$, for ε and μ sufficiently small. This will give the desired result. Now for every vector in the cone $B = \{\varepsilon\hat{\mathbf{y}} | \hat{\mathbf{y}}$ in the convex hull of $\hat{\mathbf{z}}_1, \hat{\mathbf{z}}_2, \ldots, \hat{\mathbf{z}}_{n+1}, 0 \le \varepsilon \le \varepsilon_0\}$, there is an associated attainable state $\hat{\mathbf{x}}_\varepsilon[\tau]$:

$$\sum c^i \hat{\mathbf{z}}_i \mapsto \hat{\mathbf{x}}_\varepsilon[\tau] - \hat{\mathbf{x}}[\tau] \stackrel{\text{def}}{=} \hat{\mathbf{h}}(\hat{\mathbf{c}})$$

where $\hat{\mathbf{x}}_\varepsilon[\cdot]$ is generated by the perturbations $\hat{\boldsymbol{\zeta}}_i$ at time τ_i with parameters k^i (arbitrary, but small) and c^i. This defines a continuous map $\hat{\mathbf{h}}$ of B into R^{n+1}.

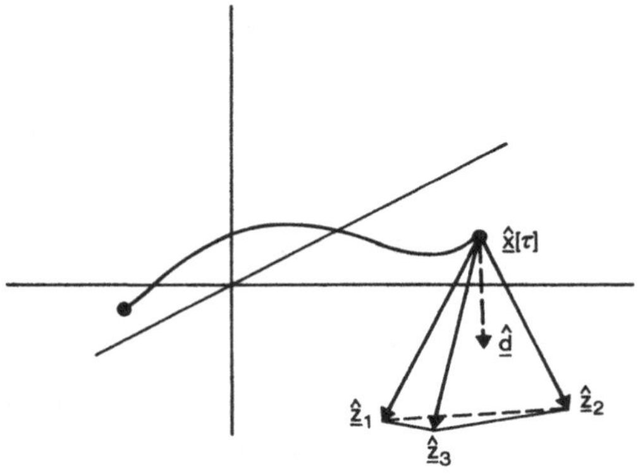

Figure 3 The Cone B

Notice that for $\hat{\mathbf{y}} \in \partial B$,

$$\|\hat{\mathbf{y}} - \hat{\mathbf{d}}\| \ge \nu > 0.$$

Thus for ε sufficiently small and $\hat{\mathbf{y}} \in \partial B$,

$$\|\hat{\mathbf{h}}(\hat{\mathbf{y}}) - \hat{\mathbf{y}}\| = \|\hat{\mathbf{x}}_\varepsilon[\tau] - \hat{\mathbf{x}}[\tau] - \varepsilon \sum c^i \mathbf{z}^i\| = o(\varepsilon) < \|\hat{\mathbf{y}} - \hat{\mathbf{d}}\|.$$

A standard variant of the Brouwer Fixed-Point Theorem (Lee and Markus [1967], p. 251) implies that $\hat{\mathbf{h}}(B)$ covers $\hat{\mathbf{d}}$. Thus there is a response $\hat{\mathbf{y}}[\cdot]$ such that $\hat{\mathbf{y}}[\tau] - \hat{\mathbf{x}}[\tau] = \hat{\mathbf{d}}$. □

Remark. The time t_1 may not be a regular point, but if $\mathcal{K}(t_1)$ is of dimension $(n+1)$ and contains $\hat{\mathbf{d}}$, then $\mathcal{K}(\tau)$ will have the same property (using, say, $\frac{1}{2}\hat{\mathbf{d}}$ in place of $\hat{\mathbf{d}}$) for τ near t_1. Since almost every point is regular, some such τ will be regular, and we can apply the lemma to conclude $(\hat{\mathbf{x}}[\cdot], \mathbf{u}(\cdot))$ is not optimal.

The proof that $\mathcal{H}(\hat{\mathbf{w}}(t), \hat{\mathbf{x}}_*[t], \mathbf{u}_*(t)) \equiv \mathcal{M}(\hat{\mathbf{w}}(t), \hat{\mathbf{x}}_*[t])$, a.e. now follows easily, as outlined earlier.

To show that $\mathcal{M} \equiv 0$, we need to enlarge our cone $\mathcal{K}(t_1)$ to a new cone $\bar{\mathcal{K}}(t_1)$ by using additional perturbations. We define (Figure 4)

$$\bar{\mathcal{K}}(t_1) = \mathcal{K}(t_1) \oplus \hat{\mathbf{f}}(\mathbf{x}_*[t_1], \mathbf{u}_*(t_1)),$$

where $A \oplus B = \{\hat{\mathbf{a}} + \beta\hat{\mathbf{b}} \| \beta| < \beta_0, \hat{\mathbf{a}} \in A, \hat{\mathbf{b}} \in B\}$.

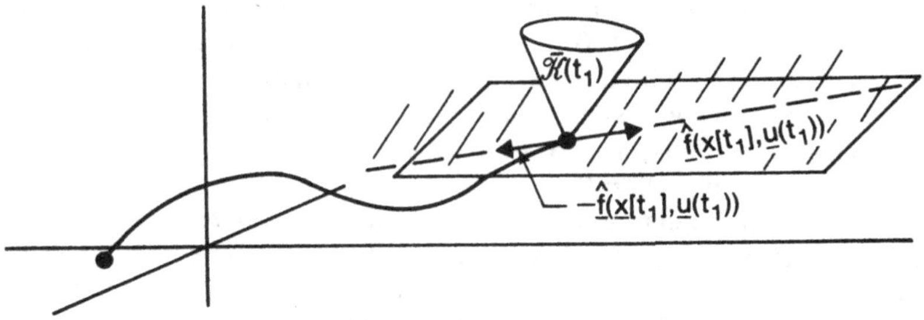

Figure 4 $\bar{\mathcal{K}}(t_1)$ Is the Shaded Half-Space

Lemma 4. *Assume that t_1 is regular. Then for β_0 sufficiently small, each $\hat{\mathbf{y}}$ in $\bar{\mathcal{K}}(t_1)$ represents an approximation to a reachable state at time t_1, in the sense that there is a response $\hat{\mathbf{x}}_\varepsilon[t]$ such that*

$$\hat{\mathbf{x}}_\varepsilon[t_1] = \hat{\mathbf{x}}_*[t_1] + \varepsilon\hat{\mathbf{y}} + o(\varepsilon).$$

Proof. We need only show that vectors of the form $\beta\hat{\mathbf{y}} \equiv \beta\hat{\mathbf{f}}(\mathbf{x}_*[t_1], \mathbf{u}_*(t_1))$ for β sufficiently small are attainable in the sense described. We will use formula (1) of Lemma 1. To get such a perturbation with $\beta > 0$, we use $\mathbf{u}_*(t + \varepsilon(c+k))$, starting at $t_0 = -\varepsilon(c+k)$ rather than $t_0 = 0$. Because the system is autonomous, this time-shift will bring us to the target $\mathbf{x}_1$ at time $t_1 - \varepsilon(c+k)$. We now leave $\mathbf{u}_\varepsilon(t)$ constant at $\mathbf{u}_*(t_1)$ for $t_1 - \varepsilon(c+k) \le t \le t_1$. Then by (1) (applied to $\hat{\mathbf{x}}_\varepsilon[\cdot]$ rather than $\hat{\mathbf{x}}_*[\cdot]$),

$$\begin{aligned}\hat{\mathbf{x}}_\varepsilon[t_1] &= \hat{\mathbf{x}}_\varepsilon[t - \varepsilon(c+k)] + \varepsilon(c+k)\hat{\mathbf{f}}(\mathbf{x}_\varepsilon[t_1], \mathbf{u}(t_1)) + o(\varepsilon)\\ &= \hat{\mathbf{x}}_1 + \varepsilon\beta\hat{\mathbf{f}}(\mathbf{x}_1, \mathbf{u}_*(t_1)) + o(\varepsilon).\end{aligned}$$

To get the final equality we used the continuity of $\hat{\mathbf{f}}$ to replace $\mathbf{x}_\varepsilon[t_1]$ by $\mathbf{x}_*[t_1]$.

To obtain the perturbation with opposite sign, we start *late*, with $\mathbf{u}_\varepsilon(t) = \mathbf{u}_*(t - \varepsilon(c+k))$ starting at time $t_0 = \varepsilon(c+k)$. This control will steer to $\mathbf{x}_1$ at time $t_1 + \varepsilon(c+k)$. If in formula (1) we take $t = \tau = t_1 + \varepsilon(c+k)$, $\sigma_1 = t_1$, then

$$\hat{\mathbf{x}}_1 = \hat{\mathbf{x}}_\varepsilon[t_1 + \varepsilon(c+k)] = \hat{\mathbf{x}}_\varepsilon[t_1] + \varepsilon(c+k)\hat{\mathbf{f}}(\mathbf{x}_\varepsilon[\tau], \mathbf{u}_\varepsilon(\tau)) + o(\varepsilon),$$

which we can rewrite as

$$\hat{\mathbf{x}}_\varepsilon[t_1] = \hat{\mathbf{x}}_1 - \varepsilon(c+k)\hat{\mathbf{f}}(\mathbf{x}_1, \mathbf{u}_*(t_1)) + o(\varepsilon). \qquad \square$$

Lemma 5

$$\mathcal{M}(\hat{\mathbf{w}}(t), \hat{\mathbf{x}}_*[t]) \equiv 0.$$

Proof. If t_1 is a regular point, then Lemma 4 shows that the vector $\beta\hat{\mathbf{f}}(\mathbf{x}_*[t_1], \mathbf{u}_*(t_1))$ is within $o(\varepsilon)$ of being in the reachable set. Lemma 3 carries over verbatim with $\mathcal{K}(t_1)$ replaced by $\mathcal{K}(t_1)$. Thus if $(\hat{\mathbf{x}}_*[\cdot], \bar{\mathbf{u}}_*(\cdot))$ is optimal, then either dim $\bar{\mathcal{K}}(t_1) < n+1$ and/or $\bar{\mathcal{K}}(t_1)$ does not contain any vertical downward vector $\hat{\mathbf{d}} = \mu(1, 0, \ldots, 0)^T$. Therefore there is a hyperplane at the vertex of $\bar{\mathcal{K}}(t_1)$ which separates it from a half-space. As before, we choose a vector $\hat{\mathbf{w}}_1$ such that $\langle \hat{\mathbf{w}}_1, \hat{\mathbf{z}} \rangle \le 0$ for all $\hat{\mathbf{z}} \in \bar{\mathcal{K}}(t_1)$. Since $\beta\hat{\mathbf{f}}(\mathbf{x}_*[t_1], \mathbf{u}_*(t_1)) \in \bar{\mathcal{K}}(t_1)$ for β both positive and negative, we conclude that $\langle \hat{\mathbf{w}}_1, \hat{\mathbf{f}}(\mathbf{x}_*[t_1], \mathbf{u}_*(t_1)) \rangle = 0$.

If t_1 is not a regular point, we can make the above argument at any regular point in $(0, t_1)$. In this way, we obtain a single regular point at which $\mathcal{H} = \langle \hat{\mathbf{w}}_1, \hat{\mathbf{f}} \rangle = 0$. We already know, from the discussion following the proof of Lemma 3, that at any regular point,

$$\mathcal{H}(\hat{\mathbf{w}}(t), \hat{\mathbf{x}}_*[t], \mathbf{u}_*(t)) = \mathcal{M}(\hat{\mathbf{w}}(t), \hat{\mathbf{x}}_*[t]),$$

$$\mathcal{H}(\hat{w}(t), \hat{\mathbf{x}}_*[t], \mathbf{u}_*(t)) \equiv \langle \hat{\mathbf{w}}(t), \hat{\mathbf{f}}(\mathbf{x}_*[t], \mathbf{u}_*(t_1)) \rangle = \mathcal{M}(\hat{w}(t), \hat{\mathbf{x}}_*[t]) = 0.$$

To complete the proof we will show that $m(t) \equiv M(\hat{w}(t), \hat{\mathbf{x}}_*[t])$ is absolutely continuous on $[0, t_1]$, and $\dot{m}(t) = 0$, a.e. Since $m(t) = 0$ at one point we will be able to conclude $m(t) \equiv 0$ on $[0, t_1]$.

To show that $m(\cdot)$ is AC, we need to show:

Given $\varepsilon > 0$, $\exists \delta > 0$ *such that for any finite set of intervals* $[t_i, t_i']$, $i = 1, \ldots, N$,

$$\sum_{i=1}^{N} |t_i - t_i'| < \delta \Rightarrow \sum_{i=1}^{N} |m(t_i) - m(t_i')| < \varepsilon.$$

Because the inequalities are strict, we can use regular points t_i, t_i'. Then

$$\begin{aligned} m(t_i) - m(t_i') &= \mathcal{H}(\hat{\mathbf{w}}(t_i), \hat{\mathbf{x}}_*[t_i], \mathbf{u}_*(t_i)) - \mathcal{H}(\hat{\mathbf{w}}(t_i'), \hat{\mathbf{x}}_*[t_i'], \mathbf{u}_*(t_i')) \\ &\ge \mathcal{H}(\hat{\mathbf{w}}(t_i), \hat{\mathbf{x}}_*[t_i], \mathbf{u}_*(t_i')) - \mathcal{H}(\hat{\mathbf{w}}(t_i'), \hat{\mathbf{x}}[t_i'], \mathbf{u}_*(t_i')) \\ &= \langle \hat{\mathbf{w}}(t_i), \hat{\mathbf{f}}(\mathbf{x}_*[t_i], \mathbf{u}_*(t_i')) \rangle - \langle \hat{\mathbf{w}}(t_i'), \hat{\mathbf{f}}(\mathbf{x}_*[t_i'], \mathbf{u}_*(t_i')) \rangle \\ &\ge -K\{|\hat{\mathbf{w}}(t_i) - \hat{\mathbf{w}}(t_i')| + |\mathbf{x}_*[t_i] - \mathbf{x}_*[t_i']|\}, \end{aligned}$$

since $H(\hat{\mathbf{w}}, \hat{\mathbf{x}}, \mathbf{u})$ is continuously differentiable with respect to $\hat{\mathbf{w}}$, $\hat{\mathbf{x}}$. Similarly,

$$\begin{aligned} m(t_i) - m(t_i') &\le \mathcal{H}(\hat{\mathbf{w}}(t_i), \hat{\mathbf{x}}_*[t_i], \mathbf{u}_*(t_i)) - \mathcal{H}(\hat{\mathbf{w}}(t_i'), \mathbf{x}_*[t_i'], \mathbf{u}_*(t_i)) \\ &\le \mathcal{K}\{|\hat{\mathbf{w}}(t_i) - \hat{\mathbf{w}}(t_i')| + |\mathbf{x}_*[t_i] - \mathbf{x}_*[t_i']|\}. \end{aligned}$$

But $\hat{\mathbf{w}}$ and $\hat{\mathbf{w}}_*$ are absolutely continuous, and the absolute continuity of $m(\cdot)$ follows.

Now $m(t') = \mathcal{H}(\mathbf{w}(t'), \mathbf{x}_*[t'], \mathbf{u}_*(t')) \equiv \mathcal{H}(t')$ when t' is regular, and $m(t) \geq \mathcal{H}(t)$ for any t. Let us compute dm/dt at a regular point t':

$$\text{(8)} \qquad \begin{aligned} \frac{m(t)-m(t')}{t-t'} &\geq \frac{\mathcal{H}(t)-\mathcal{H}(t')}{t-t'} \quad \text{for } t-t'>0, \\ \frac{m(t)-m(t')}{t-t'} &\leq \frac{\mathcal{H}(t)-\mathcal{H}(t')}{t-t'} \quad \text{for } t-t'<0. \end{aligned}$$

We write

$$\text{(9)} \qquad \frac{\mathcal{H}(t)-\mathcal{H}(t')}{t-t'} = Z(t, t') + \frac{\mathcal{H}(\hat{\mathbf{w}}(t'), \hat{\mathbf{x}}_*[t'], \mathbf{u}_*(t)) - \mathcal{H}(\hat{\mathbf{w}}(t'), \hat{\mathbf{x}}_*[t'], \mathbf{u}_*(t'))}{t-t'}$$

where

$$\begin{aligned} (t-t')Z(t, t') = {} & [\mathcal{H}(\hat{\mathbf{w}}(t), \hat{\mathbf{x}}_*[t], \mathbf{u}_*(t)) - \mathcal{H}(\hat{\mathbf{w}}(t'), \hat{\mathbf{x}}_*[t], \mathbf{u}_*(t))] \\ & + [\mathcal{H}(\hat{\mathbf{w}}(t'), \hat{\mathbf{x}}_*[t], \mathbf{u}_*(t)) - \mathcal{H}(\hat{\mathbf{w}}(t'), \hat{\mathbf{x}}_*[t'], \mathbf{u}_*(t))]. \end{aligned}$$

From (9) and the fact that $\mathbf{v} = \mathbf{u}_*(t')$ maximizes $\mathcal{H}(\hat{\mathbf{w}}(t'), \hat{\mathbf{x}}_*[t'], \mathbf{v})$, we see that

$$\text{(10)} \qquad \begin{aligned} \frac{\mathcal{H}(t)-\mathcal{H}(t')}{t-t'} &\geq Z(t, t') \quad \text{if } t-t'<0, \\ \frac{\mathcal{H}(t)-\mathcal{H}(t')}{t-t'} &\leq Z(t, t') \quad \text{if } t-t'>0. \end{aligned}$$

A straightforward application of the chain rule shows that (recall that $\langle \hat{\mathbf{w}}, \hat{\mathbf{f}} \rangle = \hat{\mathbf{w}}^T \hat{\mathbf{f}} = \hat{\mathbf{f}}^T \hat{\mathbf{w}}$)

$$\begin{aligned} \lim_{t \to t'} Z(t, t') &= \left\langle \frac{\partial \mathcal{H}}{\partial \hat{\mathbf{w}}}, \dot{\hat{\mathbf{w}}} \right\rangle + \left\langle \frac{\partial \mathcal{H}}{\partial \hat{\mathbf{x}}}, \dot{\hat{\mathbf{x}}} \right\rangle \\ &= \langle \hat{\mathbf{f}}, -[\hat{\mathbf{f}}_{\hat{\mathbf{x}}}]^T \hat{\mathbf{w}} \rangle + \langle [\hat{\mathbf{f}}_{\hat{\mathbf{x}}}]^T \hat{\mathbf{w}}, \hat{\mathbf{f}} \rangle = 0. \end{aligned}$$

Therefore, letting $t \to t'$ from the left and right, respectively, we see from (8) and (10) that

$$\frac{dm}{dt} \geq 0, \qquad \frac{dm}{dt} \leq 0,$$

respectively, for $t = t'$. Thus $dm/dt = 0$ at any regular point. □

In conclusion, we very briefly describe alternate methods of proof for the PMP. One can interpret the control problem as an abstract minimization problem, $C[\mathbf{u}(\cdot)] = \int_0^{t_1} f^0(\mathbf{x}[s], \mathbf{u}(s))\, ds$, on a function space, with constraints. Among the constraints are $\dot{\mathbf{x}} = \mathbf{f}(\mathbf{x}, \mathbf{u})$, and $\mathbf{x}[t_1] = \mathbf{x}_1$. One then obtains an abstract Lagrange multiplier problem; the Lagrange multiplier is exactly the costate $\hat{\mathbf{w}}[\cdot]$. This approach is used in Neustadt [1976], for example.

There is a very nice treatment in Craven [1978]. The advantage of this approach is that it extends quite readily to problems involving integro-differential equations, equations with delay, and so on. Its disadvantage is that it is based on a considerable amount of sophisticated mathematics, and there are subtleties which can lead the beginner into mistakes.

Another approach is that of convex analysis. This type of analysis tends to get very powerful results under minimal assumptions. This approach has led to one of the most general formulations of the PMP, with very weak hypotheses – cf. F. Clarke [1976]. Convex analysis also tends to cover a broad range of problems in optimization.

Mathematical Appendix

Control theory is an area of application of concepts from linear algebra, convexity, classical analysis, functional analysis, and differential equations. We present a selection of relevant information from each of these areas. We do not try to "survey the field," but instead concentrate on a direct, intuitive presentation of those facts which may not be familiar to all readers. In certain places we assume a familiarity with Lebesgue measure and the concept of a normed linear space.

1. Linear Algebra (References: Finkbeiner [1978], Gantmacher [1964].)

Let R denote the real numbers, let $\mathbb{C}$ denote the complex numbers, and let R^n, $\mathbb{C}^n$ denote the corresponding spaces of n-tuples.

If $A = [a_{ij}]$ is an $n \times n$ matrix with entries from $\mathbb{C}$, then a column vector $\mathbf{x} \in \mathbb{C}^n$, $\mathbf{x} \neq \mathbf{0}$, is called an *eigenvector* of A if there exists a $\lambda \in \mathbb{C}$ such that $A\mathbf{x} = \lambda \mathbf{x}$. The number λ is called an *eigenvalue*. Even if the entries of A are real, some or all of its eigenvalues and eigenvectors may be complex. For a given eigenvalue λ, the associated eigenvectors form a vector space, called the *eigenspace*. The eigenvalues of A coincide with the roots of the *characteristic equation* $\mathcal{P}_A(\lambda) \equiv \det(\lambda I - A) = 0$, where "det" stands for determinant, and I stands for the $n \times n$ identity matrix. The characteristic polynomial $\mathcal{P}_A(\lambda)$ has degree n, and we can factor $\mathcal{P}_A(\lambda)$ over $\mathbb{C}$,

$$\mathcal{P}_A(\lambda) = (\lambda - \lambda_1)^{\nu_1}(\lambda - \lambda_2)^{\nu_2} \cdots (\lambda - \lambda_p)^{\nu_p}.$$

The (complex) numbers $\lambda_1, \ldots, \lambda_p (1 \leq p \leq n)$ are the distinct eigenvalues of A, and the (natural) number ν_j is the *algebraic multiplicity* of λ_j. The dimension d_j of the eigenspace of λ_j is called the *geometric multiplicity* of λ_j, and $d_j \leq \nu_j$. An $n \times n$ matrix A always satisfies its own characteristic equation, i.e., if we replace λ in $\mathcal{P}_A(\lambda)$ by the matrix A, the result is the zero matrix, $\mathcal{P}_A(A) = 0$. We write 0 for the real or complex number zero, $\mathbf{0}$ for a null vector, and 0_{mn} for a matrix with all entries zero. When the context makes the meaning clear, we will write 0 for 0_{mn}.

Given an $n \times n$ matrix A whose distinct (complex) eigenvalues are $\{\lambda_1, \lambda_2, \ldots, \lambda_p\}$, there exists a nonsingular matrix P such that $P^{-1}AP$ is in *Jordan canonical form*, i.e., $P^{-1}AP = \mathcal{J}$ where $\mathcal{J}$ is block diagonal, i.e.,

$$(*) \qquad \mathcal{J} = \begin{bmatrix} J_1 & & 0 \\ & J_2 & \\ & & \ddots \\ 0 & & & J_k \end{bmatrix}, \qquad p \leq k \leq n;$$

each matrix J_l is square and has the form

$$J_l = \begin{bmatrix} \lambda_i & 1 & 0 \\ & \lambda_i & \ddots 1 \\ 0 & & \ddots \lambda_i \end{bmatrix} \quad \text{for some eigenvalue } \lambda_i \text{ of } A.$$

The matrices J_l, $l = 1, 2, \ldots, k$, can vary in size from 1×1 to $n \times n$, the same eigenvalue may appear in several J_l's, and each eigenvalue appears in at least one J_l.

If A is an $n \times n$ matrix over R, that is, all entries $a_{ij} \in R$, it might be the case that some entries in the matrix P and/or some eigenvalues of A are complex. We have no control over the eigenvalues of a given matrix A, but if we wish to avoid using complex entries in P, we can use the *real canonical form* of A:

> *There exists a* real *nonsingular matrix P such that $\mathcal{J} = P^{-1}AP$ is block diagonal, exactly as in* (*) *above, except the* form *of J_l is different. If the corresponding eigenvalue is real, then*
>
> $$J_l = \begin{bmatrix} \lambda_k & & 0 \\ 1 & \lambda_k & \\ 0 & \ddots 1 & \ddots \lambda_k \end{bmatrix}$$
>
> *If the corresponding* $\lambda_k = \alpha_k + i\beta_k$ (α_k, β_k real, $i = \sqrt{-1}$) *is complex, then*
>
> $$J_l = \begin{bmatrix} R_k & & \\ E_2 & R_k & \\ 0 & \ddots E_2 & \ddots R_k \end{bmatrix}, \qquad E_2 = \begin{bmatrix} 1 & 0 \\ 0 & 1 \end{bmatrix}, \qquad R_k = \begin{bmatrix} \alpha_k & -\beta_k \\ \beta_k & \alpha_k \end{bmatrix}.$$

If $A = [a_{ij}]$ is any $(m \times n)$ matrix, then $A^T = [a_{ji}]$ is its $n \times m$ transpose. If $\mathbf{x}$, $\mathbf{y}$ are column vectors in R^n, $\mathbf{x}^T = (x^1, x^2, \ldots, x^n)$, $\mathbf{y}^T = (y^1, y^2, \ldots, y^n)$, then

$$\langle \mathbf{x}, \mathbf{y} \rangle \equiv \mathbf{x}^T \mathbf{y} = \sum_{j=1}^{n} x^j y^j, \qquad |\mathbf{x}| = \sum_{j=1}^{n} |x^j|, \qquad \|\mathbf{x}\| = \langle \mathbf{x}, \mathbf{x} \rangle^{1/2}.$$

We say $\mathbf{x}$ is orthogonal (perpendicular) to $\mathbf{y}$ ($\mathbf{x} \perp \mathbf{y}$) if $\langle \mathbf{x}, \mathbf{y} \rangle = 0$.

If $\mathcal{S}=\{\mathbf{x}_1, \mathbf{x}_2, \ldots, \mathbf{x}_r\}$ is a set of vectors in R^n, then $\mathcal{S}$ is linearly independent over R if only the trivial linear combination can be zero:

$$\sum_{j=1}^{r} \alpha_j \mathbf{x}_j = \mathbf{0}, \qquad \alpha_j \in R \quad \text{for } j=1,\ldots,r \Rightarrow \alpha_1=\alpha_2=\cdots=\alpha_r=0.$$

If there is a nontrivial linear combination which yields the vector $\mathbf{0}$, then the set $\mathcal{S}$ is *dependent*. If a set $\mathcal{S}$ is dependent, then there is a nontrivial vector $\mathbf{y}\perp\mathcal{S}$, i.e., $\langle \mathbf{y}, \mathbf{x}_j\rangle = 0$ for $j=1,2,\ldots,r$. $\mathcal{S}$ is a maximal independent set in R^n if and only if it is a *basis* for R^n – every vector $\mathbf{y}\in R^n$ can be written in a unique way ($\mathbf{y}=\sum_{k=1}^{n} \alpha_k \mathbf{x}_k$) as a linear combination of the vector in $\mathcal{S}$. Every basis for R^n contains exactly n vectors.

For any set $\mathcal{S}$ of vectors in R^n and any fixed vectors $\mathbf{x}_0 \in R^n$, we define the *translate of* $\mathcal{S}$ *by* $\mathbf{x}_0$:

$$\mathbf{x}_0+\mathcal{S}=\{\mathbf{x}_0+\mathbf{y} \mid \mathbf{y}\in\mathcal{S}\}.$$

If $\mathbf{x}\in R^n$, $\mathbf{y}\in R^n$, then the Cauchy–Bunyakovskii–Schwarz inequality states

$$|\langle \mathbf{x}, \mathbf{y}\rangle| \le \|\mathbf{x}\|\,\|\mathbf{y}\|,$$

where $|\alpha|$ denotes the absolute value of the real number α. The *matrix norm* of any matrix A is

$$|A| = \sum_{i,j} |a_{ij}|.$$

There will only be a few instances where we will need to work with n-tuples of complex numbers, that is $\mathbf{x}=(x^1,\ldots,x^n)^T \in \mathbb{C}^n$. The only changes in the above are: (1) to define $\langle \mathbf{x}, \mathbf{y}\rangle \equiv \mathbf{x}^T\mathbf{y}^* \equiv \sum_{j=1}^{n} x^j (y^j)^*$, where $(\ )^*$ denotes complex conjugation; (2) to define $|\alpha+i\beta|$ as the modulus $(\alpha^2+\beta^2)^{1/2}$; and (3) to allow the α_j's to be complex in the definition of independence.

If B is an $(m\times n)$ matrix, then its *rank*, rank B, is the number of vectors in any maximal set of linearly independent row vectors of B; rank B is also the maximal number of linearly independent column vectors in B. If rank $B=r$ then there is a number $\varepsilon>0$ (depending on B) such that

$$|A-B|<\varepsilon \Rightarrow \operatorname{rank} A \ge r.$$

Finally, rank $B=r$ implies that at least one $(r\times r)$ submatrix C_r satisfies $\det C_r \ne 0$, where the $(r\times r)$ matrices C_r are formed by choosing any r rows $i_1, i_2, \ldots, i_r$ and any r columns, $j_1, \ldots, j_r$, of B and forming C_r from the entries $b_{i_k j_l}$, $k=1,\ldots,r$, $l=1,\ldots,r$.

2. Topology, Convexity, Hyperplanes (References: Eggleston [1958], Valentine [1964].)

Given $\mathbf{x}\in R^n$ and $\delta>0$, we define the *ball* of radius δ centered at $\mathbf{x}$ by $\mathcal{B}(\mathbf{x};\delta)=\{\mathbf{y}\mid \mathbf{y}\in R^n, \|\mathbf{y}-\mathbf{x}\|<\delta\}$. If $\mathcal{S}\subset R^n$ is a subset of R^n, then the

n-dimensional interior of $\mathscr{S}$ is defined

$$\text{Int } \mathscr{S} = \{\mathbf{x} \in \mathscr{S}, \exists \delta > 0 \text{ such that } B(\mathbf{x}; \delta) \subset \mathscr{S}\} \qquad (\delta \text{ may depend on } \mathbf{x}).$$

$\mathscr{S}$ is *symmetric* means $\mathbf{x} \in \mathscr{S} \Leftrightarrow -\mathbf{x} \in \mathscr{S}$. $\mathscr{S}$ is *open* if every point of $\mathscr{S}$ is in Int $\mathscr{S}$.

If $\mathscr{K}$ is a set in R^n, $\mathscr{K}^c$ will denote its complement. $\mathscr{K}$ is *closed* means $\mathscr{K}^c$ is open.

If $\mathscr{K} \subset R^n$ is a set, the *boundary* of $\mathscr{K}$ is defined by

$$\partial \mathscr{K} = \{\mathbf{x} | \mathbf{x} \notin \text{Int } \mathscr{K}, \mathbf{x} \notin \text{Int } \mathscr{K}^c\},$$

and the closure of $\mathscr{K}$ is $cl\mathscr{K} = \mathscr{K} \cup \partial \mathscr{K}$.

A set $\mathscr{K} \subset R^n$ is *convex* if the entire line segment between any two points of $\mathscr{K}$ is contained in $\mathscr{K}$, i.e.,

$$\mathbf{x}, \mathbf{y} \quad \text{in} \quad \mathscr{K} \Rightarrow \{\alpha \mathbf{x} + (1 - \alpha)\mathbf{y} | 0 < \alpha < 1\} \subset \mathscr{K}.$$

$\mathscr{K}$ is *strictly* convex if

$$\mathbf{x}, \mathbf{y} \quad \text{in} \quad \mathscr{K} \Rightarrow \{\alpha \mathbf{x} + (1 - \alpha)\mathbf{y} | 0 < \alpha < 1\} \subset \text{Int } \mathscr{K}.$$

Intuitively, the line segment between any two points of $\mathscr{K}$ lies in the interior of $\mathscr{K}$ (except perhaps for its endpoints).

Convex, but not strictly — Convex, but not strictly — Strictly convex (no line segments on the boundary)

Given a finite set of points $\{\mathbf{x}_1, \ldots, \mathbf{x}_p\} \subset R^n$, any point of the form $\sum_{j=1}^{p} \alpha_j \mathbf{x}_j$, with $\alpha_j \geq 0$ for $j = 1, 2, \ldots, p$ and $\sum_{j=1}^{n} \alpha_j = 1$, is called a *convex* combination of the points $\mathbf{x}_1, \ldots, \mathbf{x}_p$. Given a (finite or infinite) set of points $\mathscr{K} \subset R^n$, the *convex hull* of $\mathscr{K}$, co $\mathscr{K}$, is the smallest convex set containing $\mathscr{K}$. It is the intersection of all convex sets that contain $\mathscr{K}$, and is also equal to the set of *all convex combinations* of finite numbers of points in $\mathscr{K}$.

$\mathscr{K}$ — co $\mathscr{K}$ — $\mathscr{J}$ — co $\mathscr{J}$

The convex hull of a finite set of points is called a *convex polytope* (e.g., co $\mathscr{K}$ above).

A *subspace* of R^n is any set of vectors closed under addition and multiplication by scalars. Every subspace is the solution set of a homogeneous system of equations, $A\mathbf{x}=\mathbf{0}$. In R^3 a subspace is either a plane or a line through the origin. A *translate* of a subspace $\mathscr{S}$ is $\mathscr{S}+\mathbf{c}$ for some $\mathbf{c}\in R^n$, i.e., $\{\mathbf{x}|\mathbf{x}=\mathbf{y}+\mathbf{c},\ \mathbf{y}\in\mathscr{S}\}$. For example, the set $\{\mathbf{x}|2x^1-x^2=10\}$ is the translate by $\mathbf{c}=(4,-2)^T$ of the subspace $\mathscr{S}=\{\mathbf{x}|2x^1-x^2=0\}$. (Sketch it!)

An $(n-1)$-*dimensional hyperplane* P in R^n is the solution set of a single linear equation

$$P=\{\mathbf{x}|\mathbf{x}\in R^n,\ \langle\mathbf{a},\mathbf{x}\rangle=\alpha\}$$

where $\mathbf{a}$ is a given vector in R^n, and α is a given real number. It is an *affine space*, i.e., a translate of the subspace $\{\mathbf{x}|\langle\mathbf{a},\mathbf{x}\rangle=0\}$. This affine space is indeed a subspace of dimension $n-1$, and its normal is parallel to $\mathbf{a}$.

An $(n-1)$-dimensional hyperplane P defines two sets which are called *half-spaces* (they are not, unfortunately, subspaces),

$$\{\mathbf{x}|\mathbf{x}\in R^n,\ \langle\mathbf{a},\mathbf{x}\rangle-\alpha\le 0\},\qquad \{\mathbf{x}|\mathbf{x}\in R^n,\ \langle\mathbf{a},\mathbf{x}\rangle-\alpha\ge 0\}$$

and P is said to separate these sets. Given any two sets $\mathscr{S}$ and $\mathscr{K}$ in R^n, P separates them if $\mathscr{S}$ lies in one half-space determined by P, and $\mathscr{K}$ lies in the other. Two bounded closed disjoint convex sets in R^n can be separated by some hyperplane.

For a set $\mathscr{S}$, the *carrier plane* of $\mathscr{S}$ is the affine space of lowest dimension which contains $\mathscr{S}$. For example, the segment $\mathscr{S}=\{(2,y)|-1\le y\le 1\}$ has the affine space $2+\mathscr{G}$ as a carrier, where $\mathscr{G}=\{(0,y)|y\in R\}$. *When discussing* $\partial\mathscr{S}$, Int $\mathscr{S}$, etc., *we are always referring to the relative topology in the carrier plane of* $\mathscr{S}$. For the above example, $\partial\mathscr{S}=(2,-1),(2,1)$, and Int $\mathscr{S}=\{(2,y)|-1<y<1\}$. If we were to use the topology of R^2 in which $\mathscr{S}$ lies, then the boundary of $\mathscr{S}$ would be $\mathscr{S}$ itself, and $\mathscr{S}$ would have no interior. Intuitively, we work in the natural dimension of a given set rather than in the dimension of the space in which the set is given.

Let $\mathscr{K}\subset R^n$ be *convex* and *closed*. An $(n-1)$-dimensional hyperplane P supports $\mathscr{K}$ at $\mathbf{p}\in\partial\mathscr{K}\cap P$ if $\mathscr{K}$ lies entirely in one half space defined by P.

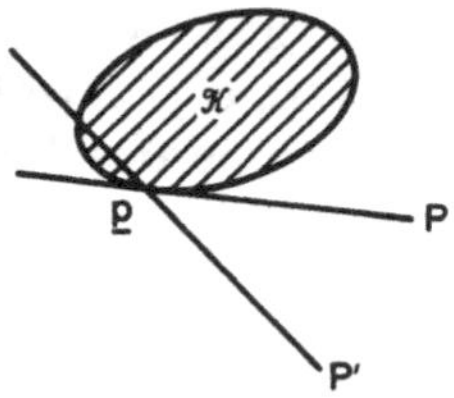

P supports $\mathscr{K}$ at $\mathbf{p}$
P' does not support $\mathscr{K}$ at $\mathbf{p}$

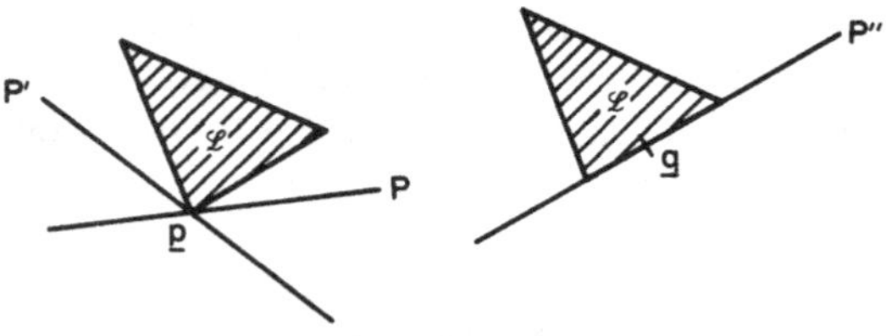

P and P' *both* support $\mathscr{L}$ at $\mathbf{p}$
P' supports $\mathscr{L}$ at $\mathbf{q}$

Through each $\mathbf{p}$ on the boundary of a *closed convex* set there passes at least one supporting $(n-1)$-dimensional hyperplane.

A point $\mathbf{p}$ is an *extreme point of a convex set* $\mathcal{K} \subset R^n$ if $\mathbf{p}$ does *not* lie on the line segment between any $\mathbf{x}$, $\mathbf{y}$ in $\mathcal{K}$ with *both* $\mathbf{x} \neq \mathbf{p}$, $\mathbf{y} \neq \mathbf{p}$. Such a point necessarily lies on $\partial\mathcal{K}$. A convex set $\mathcal{K}$ is the convex hull of its extreme points. Every supporting $(n-1)$-dimensional hyperplane of a compact convex set must contain at least one extreme point.

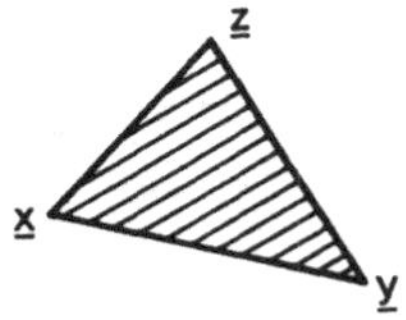

Extreme points are labelled $\mathbf{x}$, $\mathbf{y}$, $\mathbf{z}$

All boundary points except the (open) top line are extreme

All boundary points are extreme

If $\mathcal{K}$ is *closed* and *strictly convex* then each supporting $(n-1)$-dimensional hyperplane meets $\mathcal{K}$ in *exactly one* point. A closed strictly convex set containing more than one point must have a nonempty interior, and every boundary point is extreme.

If $\mathcal{K}$ is convex and *compact* (closed and bounded) in R^n, and $\mathbf{y} \notin \mathcal{K}$, then $\mathbf{y}$ and $\mathcal{K}$ can be *strictly* separated by a hyperplane $\mathbf{P} = \{\mathbf{z} | \langle \mathbf{z}, \mathbf{a} \rangle = \alpha\}$: $\langle \mathbf{a}, \mathbf{x} \rangle \leq \alpha$ for all $\mathbf{x} \in \mathcal{K}$, and $\langle \mathbf{a}, \mathbf{y} \rangle > \alpha$.

Finally we need to establish one special, but simple, result for convex sets. Let $\mathcal{K}$ be convex and $\mathbf{p} \in \partial\mathcal{K}$. Then there is an $(n-1)$-dimensional supporting hyperplane $P = \{\mathbf{x} | \langle \mathbf{a}, \mathbf{x} \rangle = \alpha\}$ at $\mathbf{p}$. In particular $\langle \mathbf{a}, \mathbf{p} \rangle = \alpha$. Since $\mathcal{K}$ lies on one side of P we have either $\langle \mathbf{a}, \mathbf{q} \rangle \leq \alpha$ for $\mathbf{q} \in \mathcal{K}$, or the opposite inequality throughout $\mathcal{K}$. By using $\mathbf{b} = -\mathbf{a}$ in the latter case and $\mathbf{b} = \mathbf{a}$ in the first case, we can assert:

For each $\mathbf{p} \in \partial\mathcal{K}$ *there is a vector* $\mathbf{b} \in R^n$ *(depending on* $\mathbf{p}$*) such that*

$$\langle \boldsymbol{b}, \boldsymbol{p} \rangle = \sup_{\mathbf{q} \in \mathcal{K}} \langle \mathbf{b}, \mathbf{q} \rangle.$$

Metric and Normed Spaces, the Hausdorff Metric, Weak Compactness (References: Yosida [1966], Goffman and Pedrick [1965], Royden [1963].) A *metric space* is any collection of objects $X = \{A, B, C, \ldots\}$ with a (distance) function $\rho : X \times X \to R$ which has the following properties: For all A, B, C in X,

(i) $\rho(A, B) \geq 0$, and $\rho(A, B) = 0 \Leftrightarrow A = B$.
(ii) $\rho(A, B) = \rho(B, A)$.
(iii) $\rho(A, B) \leq \rho(A, C) + \rho(C, B)$.

R^n is a metric space with either $\rho_1(\mathbf{x}, \mathbf{y}) = |\mathbf{x} - \mathbf{y}|$ or $\rho_2(\mathbf{x}, \mathbf{y}) = \|\mathbf{x} - \mathbf{y}\|$, where as always, $|z| = \sum_1^n |z^i|$, $\|z\|^2 = \sum_1^n |z^i|^2$.

Let X be the collection of all closed subsets of R^n. For $\mathbf{x} \in R^n$, $A \in X$, $B \in X$, we define

$$d(\mathbf{x}, A) = \inf \{\|x - y\| | \mathbf{y} \in A\}, \qquad N(A; \varepsilon) = \{\mathbf{x} | \mathbf{x} \in R^n, d(\mathbf{x}, A) < \varepsilon\}.$$

Then we can define the *extended Hausdorff metric* on X:

$$h(A, B) = \inf \{\varepsilon | A \subset N(B; \varepsilon), B \subset N(A; \varepsilon)\}.$$

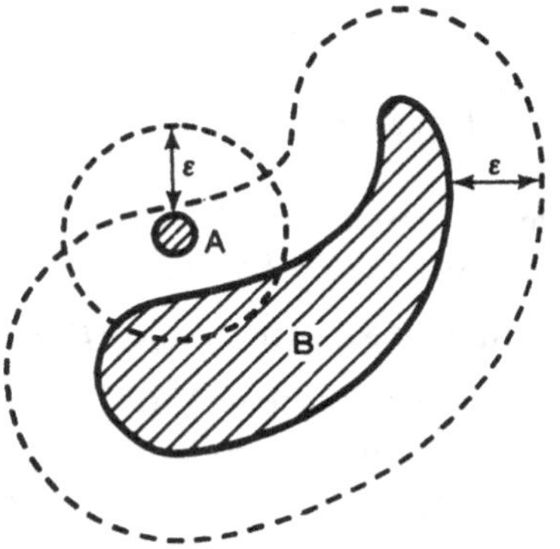

$A \subset N(B, \varepsilon)$ but $B \not\subset N(A, \varepsilon)$

For any two closed sets A, B in R^n, $h(A, B)$ is either a non-negative real number or $+\infty$. For example if A is a single point in R^2 and B is the closed half space $\{(x^1, x^2) | x^1 \leq 0\}$, then $h(A, B) = +\infty$. This extended metric will satisfy the distance axioms above if we adopt the usual convention $\infty + \infty = \infty$. If we choose X to be the closed and bounded subsets of R^n, then $h(A, B)$ will be a metric. If we were to allow the target set $\mathcal{T}(t)$ in control theory to be unbounded, then we would need the extended metric. One simple fact which we will need later for closed sets A, B in R^n is:

$$A \subset B \Rightarrow h(A, B) = \sup_{\mathbf{x} \in B} d(\mathbf{x}, A) = \sup_{\mathbf{x} \in B} \inf_{\mathbf{y} \in A} \|\mathbf{x} - \mathbf{y}\|.$$

A set S in a metric space is *sequentially compact*, if any sequence from S contains a subsequence which converges to a limit in S.

We will make considerable use of the normed linear spaces of R^n-valued functions, $L^2[0, t_1]$ and $L^1[0, t_1]$. For measurable functions $\mathbf{f}(\cdot): [0, t] \to R^n$, we define the norms as, respectively,

$$\|\mathbf{f}(\cdot)\|_2 = \left\{ \int_0^{t_1} \|\mathbf{f}(\tau)\|^2 \, d\tau \right\}^{1/2}, \qquad \|\mathbf{f}(\cdot)\|_1 = \int_0^{t_1} \|\mathbf{f}(\tau)\| \, d\tau,$$

where $\|\mathbf{f}(\tau)\| = [\mathbf{f}^T(\tau)\mathbf{f}(\tau)]^{1/2}$.

If $S \subset [0, t_1]$, then $|S|$ will denote the Lebesgue measure of S, and $\chi_S(t)$ is its characteristic function i.e., $\chi_S(t)$ is $+1$ on S, 0 elsewhere.

If X is a normed linear space (NLS), with norm $\|\cdot\|$, then for $\mathbf{x}\in X$ and $\delta>0$ we define the ball $\mathscr{B}(\mathbf{x};\delta)=\{y \mid \|\mathbf{x}-\mathbf{y}\|<\delta\}$. If $Z\subset X$, then the *closure* of Z is

$$\bar{Z}=\{\mathbf{x}\in X \mid \forall\delta>0,\ \exists\mathbf{z}\in Z,\ \mathbf{z}\in\mathscr{B}(\mathbf{x},\delta)\}.$$

If $\bar{Z}=X$ then Z is *dense* in X. We will use the fact that $\mathscr{U}_{PC}[0,t_1]$ is L^1-dense in $\mathscr{U}_m[0,t_1]$. This means that given any $\mathbf{u}(\cdot)\in\mathscr{U}_m[0,t_1]$ and any $\delta>0$, there is a $\mathbf{u}_\delta(\cdot)\in\mathscr{U}_{PC}[0,t_1]$ (i.e., a piecewise constant function) such that $\mathbf{u}_\delta(\cdot)\in\mathscr{B}(\mathbf{u}(\cdot);\delta)$. In other words,

$$\int_0^{t_1}\|\mathbf{u}_\delta(\tau)-\mathbf{u}(\tau)\|\,d\tau<\delta.$$

In fact $\mathscr{U}_{PC}[0,t_1]$ is "dense in measure" in $L^1[0,t_1]$, that is, $\mathbf{u}_\delta(\cdot)$ is "pointwise close to $\mathbf{u}(\cdot)$ most of the time": For any $\delta_1>0$, $\delta_2>0$,

$$\exists S\subset[0,t_1],\quad\text{with } |S|<\delta_1, \text{ such that } \|\mathbf{u}_\delta(t)-\mathbf{u}(t)\|<\delta_2 \text{ on } S^c.$$

For a full discussion of this, see Royden [1963].

We will need a few facts about weak convergence and weak compactness. In $L^2[0,t_1]$, a sequence $\{\mathbf{v}_k(\cdot)\ (k=1,2,\ldots)$ converges weakly to $\mathbf{v}(\cdot)$, $\mathbf{v}_k(\cdot)\overset{\omega}{\Rightarrow}\mathbf{v}(\cdot)$; if

$$\forall\mathbf{g}(\cdot)\in L^2[0,t_1],\qquad \lim_{k\to\infty}\int_0^{t_1}\mathbf{g}^T(\tau)\mathbf{v}_k(\tau)d\tau=\int_0^{t_1}\mathbf{g}^T(\tau)\mathbf{v}(\tau)\,d\tau.$$

A subset $Z\subset L^2[0,t_1]$ is *sequentially weakly compact* if any sequence $\{\mathbf{z}_n\}\subset Z$ has a weakly convergent subsequence with limit in Z. If X, Y are normed linear spaces, and $F(\cdot):X\to Y$ is such that $x_n\overset{\omega}{\Rightarrow}x_0\Rightarrow \lim_{k\to\infty}F(x_n)=F(x_0)$ then $F(\cdot)$ maps *weakly sequentially compact* sets into *sequentially compact* sets. This is an analogue of the standard result that a continuous function maps compact sets into compact sets.

In $L^2[0,t_1]$, any closed bounded set (in particular, every closed ball $\operatorname{cl}\mathscr{B}(\mathbf{f}(\cdot),\delta)$) is weakly sequentially compact.

Absolute Continuity; *Jacobians*; *the "o"-Symbol.* A function $f:R^1\to R^1$ is absolutely continuous (AC) on an interval I if for each $\varepsilon>0$ there exists a $\delta>0$ such that

$$\sum_{i=1}^N(\beta_i-\alpha_i)<\delta\Rightarrow\sum_{i=1}^N|f(\beta_i)-f(\alpha_i)|<\varepsilon,$$

whenever $(\alpha_1,\beta_1),\ldots,(\alpha_N,\beta_N)$ are disjoint subintervals of I. If $f(t)$ is AC, then $\dot{f}(t)$ exists, a.e., and $f(t)=f(a)+\int_a^t\dot{f}(s)\,ds$ whenever $[a,t]$ is in the domain of f. The extension to vector functions $\mathbf{f}(\cdot):R^1\to R^n$ is immediate, using components.

The set of continuous functions possessing all partial derivatives up to and including the kth on a set $D\subset R$ will be denoted $C^k(D;E)$, where E

is the range space (usually R^n). By a kth partial derivative of a function $\mathbf{f}(\mathbf{x})$ from R^p to R^m we mean one of:

$$\frac{\partial^k f^j}{\partial^{k_1} x^1 \partial^{k_2} x^2 \cdots \partial^{k_p} x^p}, \qquad \sum_{s=1}^{p} k_s = k; \qquad j = 1, \ldots, n.$$

The *Jacobian matrix at* $\mathbf{x}_0$ of a function $\mathbf{f}(\mathbf{x})$ in $C^1(R^p, R^q)$ is the $(q \times p)$-matrix $\mathbf{f}_\mathbf{x}(\mathbf{x}_0) \equiv [\partial f^i/\partial x^j]|_{\mathbf{x}=\mathbf{x}_0}$. If $p = q$ and $\det \mathbf{f}_\mathbf{x}(\mathbf{x}) \neq 0$ on an open set $\mathscr{S} \subset R^p$, then $\mathbf{f}(\cdot)$ is an *open* map, i.e., for any open $\mathscr{K} \subset \mathscr{S}$,

$$\mathbf{f}(\mathscr{K}) \equiv \{\mathbf{f}(\mathbf{x}) | \mathbf{x} \in \mathscr{K}\} \quad \text{is open.}$$

If $f(\mathbf{z})$ is a real or complex valued function of the vector variable $\mathbf{z}$, then the statement "$f(\mathbf{z}) = o(|\mathbf{z}|)$ as $\mathbf{z} \to 0$" means

$$\lim_{\mathbf{z}\to 0} \frac{f(\mathbf{z})}{|\mathbf{z}|} = 0.$$

For example, for x real, $f(x) = x^{4/3}$ is $o(|x|)$, $g(x) = x^{1/3}$ is not.

3. Theory of Ordinary Differential Equations (References: Coddington and Levinson [1955], Hartman [1964].)

General Existence and Continuity Theorems (Coddington and Levinson [1955], Chapters 1 and 2.) The differential equation $\dot{\mathbf{x}} = \mathbf{f}(t, \mathbf{x})$, $\mathbf{x}(t) \in R^n$, $\mathbf{f}(t, \mathbf{x}) \in R^n$, with specified initial condition $\mathbf{x}(t_0) = \mathbf{x}_0$, will have a solution in some neighborhood of t_0 if either

(a) $\mathbf{f}(t, \mathbf{x})$ is continuous in its $(n+1)$ variables or
(b) $\mathbf{f}(t, \mathbf{c})$ is a measurable function of t for any constant $\mathbf{c}$ in a neighborhood of $\mathbf{x}_0$, $\mathbf{f}(t_1, \mathbf{x})$ is a continuous function of $\mathbf{x}$ for each t_1 in a neighborhood of t_0, and $|\mathbf{f}(t, \mathbf{x})| \leq m(t)$ near $(t_0, \mathbf{x}_0)$, where $m(t)$ is integrable.

Of course, condition (a) is contained in (b), but (a) is easier for the novice to grasp. The conditions in (b) are called the Caratheodory conditions. In case (b), a solution is an absolutely continuous function which satisfies the differential equation almost everywhere.

In either case, the solution will be unique if $\mathbf{f}(t, \mathbf{x})$ satisfies a Lipschitz condition in $\mathbf{x}$:

$$|\mathbf{f}(t, \mathbf{x}_1) - \mathbf{f}(t, \mathbf{x}_2)| \leq K |\mathbf{x}_1 - \mathbf{x}_2|$$

for some constant K, all t near t_0, and all $\mathbf{x}_1$, $\mathbf{x}_2$ near $\mathbf{x}_0$. For more general uniqueness results see Hartman [1964], Chapters II and III.

The equation $\dot{x} = x^2$ with $x(0) = 1$, has the solution $x(t) = 1/(1 - t)$, which only exists on $(-\infty, 1)$. This shows that even for simple equations, extendability can be a problem.

If the solution of the initial value problem (IVP) $\dot{\mathbf{x}} = \mathbf{f}(t, \mathbf{x})$, $\mathbf{x}(t_0) = \mathbf{x}_0$ *is unique* and exists in $[a, b]$, in either case (a) or case (b), then the solution of the "nearby" IVP $\dot{\mathbf{y}} = \mathbf{f}(t, \mathbf{y})$, $\mathbf{y}(t_1) = \mathbf{x}_1$ will exist on $[a, b]$ for $(t_1, \mathbf{x}_1)$ close to $(t_0, \mathbf{x}_0)$ (it may not be unique). In addition, the solution $\mathbf{y}(t; t_1, \mathbf{x}_1)$ depends continuously on $(t_1, \mathbf{x}_1)$: as $(t_1, \mathbf{x}_1) \to (t_0, \mathbf{x}_0)$, $\mathbf{y}(t; t_1, \mathbf{x}_1) \to \mathbf{x}(t; t_0, \mathbf{x}_0)$ uniformly on $[a, b]$. Furthermore, if $\mathbf{f}(t, \mathbf{x})$ contains any parameters on which it depends continuously, then the solutions will depend continuously on these parameters. In fact, even the interval of existence of a solution will depend continuously on these parameters.

Linear Equations. If $A(t)$ is an $n \times n$ matrix and $\mathbf{b}(t) \in R^n$, for t on an interval $I \subset R$, then any vector ordinary differential equation (o.d.e.) of the form

$$\text{(L)} \qquad \dot{\mathbf{x}}(t) = A(t)\mathbf{x}(t) + \mathbf{b}(t), \qquad t \in I \subset R,$$

is a linear o.d.e. If each entry in $A(t)$, $\mathbf{b}(t)$ is an integrable function on I, then the solution of the initial value problem (L), $\mathbf{x}(t_0) = \mathbf{x}_0$ (given $t_0 \in I$, $\mathbf{x}_0 \in R^n$) is unique and exists on all of I. By a *solution* of (L) we mean an absolutely continuous function $\mathbf{x}(t)$ that satisfies (L) almost everywhere. If $A(t)$, $\mathbf{b}(t)$ are in fact continuous on I, then a solution will be differentiable throughout I and will satisfy (L) everywhere. We will assume throughout this section that $A(t)$, $\mathbf{b}(t)$ are continuous, to avoid repeating the phrase "almost everywhere." Associated with the *nonhomogeneous* system (L) is the *homogeneous* system

$$\text{(H)} \qquad \dot{\mathbf{x}}(t) = A(t)\mathbf{x}(t).$$

Let $\mathbf{x}_1(t), \mathbf{x}_2(t), \ldots, \mathbf{x}_n(t)$ be solutions of (H) on I. These functions yield linearly independent vectors at one $t_0 \in I$ if and only if they give linearly independent vectors at every $t \in I$ (t_0 is arbitrary). In this case, the set of solutions is called a fundamental solution set for (H), and the $n \times n$ nonsingular matrix function

$$X(t) = [\mathbf{x}_1(t), \mathbf{x}_2(t), \ldots, \mathbf{x}_n(t)], \qquad t \in I$$

is called a *fundamental* matrix. If $X(t)$, $Y(t)$ are two fundamental matrices for a given equation (H), then there exists a nonsingular $n \times n$ *constant* matrix C such that $X(t) = Y(t)C$. Given a fundamental matrix $X(t)$ of (H), the entire solution set of (H) is $\{X(t)\mathbf{c} \mid \mathbf{c} \in R^n\}$.

The unique solution of the nonhomogeneous initial value problem (L), $\mathbf{x}(t_0) = \mathbf{x}_0$, is given by the variation of parameters formula:

$$\mathbf{x}(t) = X(t)X^{-1}(t_0)\mathbf{x}_0 + X(t) \int_{t_0}^{t} X^{-1}(s)\mathbf{b}(s)\, ds,$$

where $X(t)$ is any fundamental matrix for (H) (we integrate matrices and vectors component-wise).

Constant Coefficients. If $A(t) \equiv A$ is a constant matrix, then one fundamental matrix for the *constant coefficient homogeneous equation*

$$\text{(CCH)} \qquad \dot{\mathbf{x}} = A\mathbf{x}$$

is the matrix exponential $X(t) = e^{At} \equiv \sum_{k=0}^{\infty} (A^k t^k / k!)$. Each entry of e^{At} is a linear combination of terms of the form $t^k e^{\lambda_j t}$, where $\{\lambda_1, \ldots, \lambda_r\}$ are the distinct eigenvalues of A, and for each j, $k = 0, 1, 2, \ldots, m_j$, with $m_j \geq 0$. In particular, an application of the theory described above shows that the general solution of the constant coefficient scalar equation, $\sum_{k=0}^{n} a_k D^k y(t) = 0$ $(D \equiv d/dt)$, is a linear combination of terms of the form $t^j e^{\lambda t}$, where λ satisfies $\sum_{k=0}^{n} a_k \lambda^k = 0$ and j is a non-negative integer.

The nonhomogeneous constant-coefficient initial-value problem

$$\text{(CCNH)} \qquad \dot{\mathbf{x}}(t) = A\mathbf{x}(t) + \mathbf{b}(t), \qquad \mathbf{x}(t_0) = \mathbf{x}_0$$

has the unique solution

$$\mathbf{x}(t) = e^{A(t-t_0)}\mathbf{x}_0 + e^{At} \int_{t_0}^{t} e^{-As}\mathbf{b}(s)\, ds.$$

Stability (References: Cesari [1963]; LaSalle and Lefschetz [1961], Hahn [1967]). Consider the initial value problem

$$\dot{\mathbf{x}}(t) = \mathbf{f}(t, \mathbf{x}(t)), \qquad \mathbf{x}(t_0) = \mathbf{x}_0, \qquad \mathbf{f} \in C([t_0, \infty) \times R^n).$$

If $\mathbf{x}[t] \equiv \mathbf{x}(t; t_0, \mathbf{x}_0)$ is a solution of this problem, then $\mathbf{x}[t]$ is said to be *Liapunov stable* (to the right) if

(1) $\mathbf{x}[t]$ extends to $[t_0, \infty)$,
(2) there is a $\delta_1 > 0$ such that all solutions $\mathbf{x}(t; t_0, \tilde{\mathbf{x}}_0) \equiv \tilde{\mathbf{x}}[t]$ with $|\mathbf{x}_0 - \tilde{\mathbf{x}}_0| < \delta$ extend to $[t_0; \infty)$;
(3) for any $\varepsilon > 0$ there is a $\delta > 0$ such that $\mathbf{x}_0 |\mathbf{x}_0 - \tilde{\mathbf{x}}_0| < \delta(\varepsilon) \Rightarrow |\mathbf{x}[t] - \tilde{\mathbf{x}}[t]| < \varepsilon$, $0 \leq t \leq +\infty$.

Briefly, $\mathbf{x}[\cdot]$ Liapunov stable means that solutions which start close to $x[t_0]$ stay close to $\mathbf{x}[\cdot]$. If in addition to (3) we require

(4) there is a $\delta > 0$ such that for all $\tilde{\mathbf{x}}_0$,

$$|\mathbf{x}_0 - \tilde{\mathbf{x}}_0| < \delta \Rightarrow \lim_{t \to \infty} |\mathbf{x}[t] - \tilde{\mathbf{x}}[t]| = 0,$$

then $\mathbf{x}[t]$ is *asymptotically stable*. This solution is *globally asymptotically stable* if it is stable and *all* solutions tend to it, i.e., we can choose $\delta = +\infty$ in (4). Any solution satisfying (4) is called an *attractor*. An attractor need not be stable, a stable solution need not be an attractor.

One useful result for simple stability conclusions is *Gronwall's Inequality*:

If $f(t)$, $g(t)$, $h(t)$ are continuous on $[a, b]$, with $h(t) \geq 0$, and if

$$x(t) \leq f(t) + g(t) \int_a^t h(s)x(s)\, ds, \qquad a \leq t \leq b$$

then

$$x(t) \leq f(t) + g(t) \int_a^t h(s)f(s)\, e^{\int_s^t h(r)g(r)\,dr}\, ds.$$

Autonomous Equations. A system is called *autonomous* if there is no explicit appearance of t in it:

(A) $$\dot{\mathbf{x}} = \mathbf{f}(\mathbf{x}).$$

Notice that if $\mathbf{x}(t)$ solves (A), then so does $\mathbf{x}(t+\alpha)$ for any $\alpha \in R$. Suppose $\mathbf{x}(t)$ solves (A) on $[t_0, t_1]$ and traces out the path sketched below.

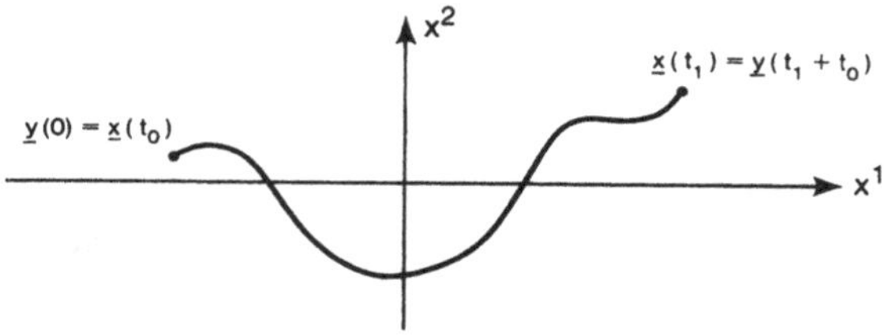

Then $\mathbf{y}(t) = \mathbf{x}(t - t_0)$ also solves (A), and traces out the same path *as t varies from* 0 *to* $t_1 + t_0$, $0 \leq t \leq t_1 + t_0$. Thus with an autonomous equation there is no loss of generality in setting $t_0 = 0$.

A vector $\mathbf{c}$ such that $\mathbf{f}(\mathbf{c}) = \mathbf{0}$ is called an *equilibrium point* (rest point, critical point) of (A), since $\mathbf{x}(t) \equiv \mathbf{c}$ is then a solution. One can then discuss the stability of this equilibrium.

Because the general solution of the linear autonomous system (LA) $\dot{\mathbf{x}} = A\mathbf{x}$, with A a constant matrix, is completely known in terms of the eigenvalues of A, the stability of the trivial solution $\mathbf{x}(t) \equiv \mathbf{0}$ of (LA) can be described in terms of these eigenvalues (below, Re z means "the real part of the complex number z"):

The zero solution of (LA) is stable if and only if: Re $\lambda_i \leq 0$ *for each eigenvalue of A, and when* Re $\lambda_k = 0$, *then λ_k only occurs in 1×1 blocks in the Jordan form of A. This last is equivalent to saying that if λ_k has multiplicity m_k in the characteristic equation of A, then corresponding to λ_k there are m_k linearly independent eigenvectors.*

The zero solution of (LA) is asymptotically *stable if and only if* Re $\lambda_i < 0$ *for each eigenvalue of A (and this solution is therefore* globally *asymptotically stable).*

An analysis of the stability of a critical point $\mathbf{c}$ for a system $\mathbf{x} = \dot{\mathbf{x}} = \mathbf{f}(\mathbf{x})$ with $\mathbf{f} \in C^2(D; R^n)$, with $\mathbf{c} \in D$ and D open, can be carried out to a certain extent by using Taylor's expansion about $\mathbf{c}$ (remember $\mathbf{f}(\mathbf{c}) = \mathbf{0}$). One then gets the *linearized* system

$$\dot{\mathbf{x}}(t) = [\partial f^i/\partial x^j]|_{\mathbf{x}=\mathbf{c}}\mathbf{x}(t) + \boldsymbol{\varepsilon}(t, \mathbf{x}),$$

where $\boldsymbol{\varepsilon}(\cdot, \cdot)$ is the remainder. Under reasonable conditions, the stability of the linear system $(\varepsilon(t, \mathbf{x}) \equiv 0)$ determines the stability of the original system.

Bibliography

1. Allen, R. G. D. [1966]: *Mathematical Economics*, Macmillan, New York.
2. Athans, M. and Falb, P. [1966]: *Optimal Control*, McGraw-Hill, New York.
3. Auslander, D., Takahashi, Y., and Rabins, M. J. [1974]: *Introducing Systems and Control*, McGraw-Hill, New York.
4. Balakrishnan, A. B. [1973]: *Stochastic Differential Systems I*, Springer-Verlag, New York.
5. Bellman, R. E. [1953]: *Stability Theory of Ordinary Differential Equations*, Academic, New York.
6. Bellman, R. E. [1967]: *Introduction to the Mathematical Theory of Control Processes*, Vol. I, Academic, New York.
7. Bellman, R. E. [1971]: *ibid*, Vol. II.
8. Bennett, S. [1979]: *A History of Control Engineering*, Peter Peregrinus, Stevenage, England.
9. Berkovitz, L. D. [1974]: *Optimal Control Theory*, Springer-Verlag, New York.
10. Berkovitz, L. D. [1978]: Existence Theory for Optimal Control Problems, appearing in Schwarzkopf, Kelly, and Eliason [1978].
11. Cesari, L. [1968]: *Asymptotic Behavior and Stability Problems in Ordinary Differential Equations*, Springer-Verlag, New York.
12. Cesari, L. [1971]: Closure, lower closure and semi-continuity theorems in optimal control, SIAM J. Control, **9**, 287–315.
13. Clark, C. W. [1976]: *Mathematical Bioeconomics*, Wiley-Interscience, New York.
14. Clarke, Frank H. [1976]: The maximum principle under minimal hypotheses, SIAM J. Control and Optimization, **14**, 1028–1091.
15. Coddington, E. A. and Levinson, N. [1955]: *Theory of Ordinary Differential Equations*, McGraw-Hill, New York.
16. Coppel, W. A. [1965]: *Stability and Asymptotic Behavior of Differential Equations*, Heath, Boston.
17. Craven, B. D. [1978]: *Mathematical Programming and Control Theory*, Chapman and Hall, London.
18. Eggleston, H. [1958]: *Convexity*, Cambridge University Press, Cambridge, England.

19. Fillipov, A. F. [1959]: On certain questions in the theory of optimal control, SIAM J. Control, **1**, 76–84 .
20. Finkbeiner, D. [1978]: *An Introduction to the Theory of Matrices and Linear Transformations*, Third Edition, W. H. Freeman, San Francisco.
21. Fleming, W. H. and Rishel, R. W. [1975]: *Deterministic and Stochastic Control Theory*, Springer-Verlag, New York.
22. Gantmacher, F. R. [1964]: *Matrix Theory*, Chelsea, New York.
23. Goffman, C. and Pedrick, G. [1965]: *First Course in Functional Analysis*, Prentice-Hall, Englewood Cliffs, NJ.
24. Hadley, G. and Kemp, M. C. [1971]: *Variational Methods in Economics*, North-Holland, New York.
25. Hahn, W. [1967]: *Stability of Motion*, Springer-Verlag, New York.
26. Hartman, P. [1964]: *Ordinary Differential Equations*, Wiley, New York.
27. Hermes, H. and LaSalle, J. P. [1969]: *Functional Analysis and Time-Optimal Control*, Academic, New York.
28. Hestenes, M. R. [1966]: *Calculus of Variations and Optimal Control*, Wiley, New York.
29. Intriligator, M. D. [1971]: *Mathematical Optimization and Economic Theory*, Prentice-Hall, Englewood Cliffs, NJ.
30. Isaacs, R. I. [1975]: *Differential Games*, Krieger, Huntington, New York.
31. Jacobs, M. Q. [1968]: Attainable sets in systems with unbounded controls, J. Diff. Eq., **4**, 408–423.
32. LaSalle, J. P. and Lefschetz, S. [1961]: *Stability by Liapunov's Direct Method with Applications*, Academic, New York.
33. Lee, E. B. and Markus, L. [1967]: *Foundations of Optimal Control Theory*, Wiley, New York.
34. Leitmann, G. (editor) [1967]: *Topics in Optimization*, Academic, New York.
35. Mayr, Otto [1970]: The origins of feedback control, Scientific American, October, 1970, 110–118.
36. McShane, E. J. [1978]: The Calculus of Variations From the Beginning Through Optimal Control Theory, appearing in Schwarzkopf, Kelly, and Eliason [1978].
37. Milhorn, H. T. [1966]: *The Application of Control Theory to Physiological Systems*, Saunders, Philadelphia.
38. Neustadt, L. [1976]: *Optimization: A Theory of Necessary Conditions*, Princeton University Press, Princeton, NJ.
39. Oguztoreli, M. N. [1966]: *Time Lag Control Systems*, Academic, New York.
40. Olech, C. [1966]: Extremal solutions of a control system, J. Diff. Eq., **2**, 74–101.
41. Pontryagin, L. S., Boltyanskii, V. G., Gamkrelidze, R. S. and Mishchenko, E. F. [1964]: *The Mathematical Theory of Optimal Processes*, Pergamon–Macmillan, New York.
42. Ramsey, F. P. [1928]: A mathematical theory of saving, Quarterly Econ. J., **38**, 543–549.
43. Royden, H. L. [1963]: *Real Analysis*, Macmillan, New York.
44. Roxin, E. [1962]: The existence of optimal controls, Michigan Math. J., **9**, 109–119.
45. Russell, D. L. [1964]: Penalty functions and bounded phase coordinate control, SIAM J. Control, **2**, 409–422.
46. Sansone, G. and Conti, R. [1964]: *Nonlinear Differential Equations*, Macmillan, New York.
47. Schwarzkopf, A. B., Kelley, W. G. and Eliason, S. B. (editors) [1978]: *Optimal Control and Differential Equations*, Academic, New York.
48. Strauss, A. [1968]: *An Introduction to Optimal Control Theory*, Lecture Notes in Operations Research and Mathematical Economics, Vol. 3, Springer-Verlag, New York.

49. Valentine, F. A. [1964]: *Convex Sets*, McGraw-Hill, New York.
50. Waltman, P. [1974]: *Deterministic Threshold Models in the Theory of Epidemics*, Springer-Verlag, New York.
51. Yosida, K. [1966]: *Functional Analysis*, Springer-Verlag, New York.

Index

A priori bound 85
Absolute continuity 154
Adjoint system 79, 105–107
Affine system 151
Asymptotic stability 157
Attractor 157
Autonomous equation 158

Ball 149
Bang–bang control 7
Bang–bang principle 30, 50–55, 59, 97, 128
Basis 149
Bibliography 160
Boundary 150

$\mathscr{C}$ 6, 24 *et seq.*
$\mathscr{C}_{BB}$ 29, 45
$\mathscr{C}_{BBPC}$ 29, 45
$\mathscr{C}^k$ 154
$\mathscr{C}_{PC}$ 45
Calculus of variations 21
Caratheodory conditions 155
Carrier plane 151
Characteristic equation 147
Characteristic function 153
Characteristic polynomial 147
$cl(K)$ 150
Closed 150
Closure 154
Compact 151–152
Compact, sequentially 153
Completely controllable 26
Constraints 2
Control 1, 5
Control problem 5
Controllability matrix 31, 38
Controllable set 6
Controllable state 6
Convex 150
Convex combination 150
Convex hull 150
Convex polytope 150
Convex, strictly 150
Cost functional 9
Costate 105

Differential equations 21, 155 *et seq.*
Dynamic programming 21, 118
Dynamics 3, 5

Economics model 19
Economy, national 1
Eigenspace 147
Eigenvalue 147
Eigenvector 147
Epidemics 17
Equilibrium 158
Existence 82 *et seq.*
Extended velocity vector 91
Extremal control 61
Extreme point 152

Feedback 2, 6
Fillipov's lemma 94
Fundamental matrix 156

Game theory 20
Global asymptotic stability 157
Gronwall's inequality 158

$H(\mathbf{w}, \mathbf{x}, \mathbf{u})$ 108
Half space 151
Hausdorff metric 153
Homogeneous system 156
Hyperplane 151

Int S 150
Interior 150

Jacobian matrix 155
Jordan canonical form 148

$K(\mathrm{t}, \mathbf{x})$ 7
K_{BBPC} 10

Liapunov stability 157
Linear independence 149
Linearization 159
Lipschitz 155
Lipschitz condition 88

$M(\mathbf{w}, \mathbf{x})$ 108
Manifold 124
Maximum principle 9, 58, 64, 77, 79, 108, 127, 134 *et seq.*
Metric space 152
Moon landing 18
Multiplicity
  algebraic 147
  geometric 148

Necessary conditions 82
Nonhomogeneous system 156
Norm 149, 153
Normality 58, 66, 71–73, 79

$o(\mathbf{z})$ 155
Objective 2
Observable 22
Open 150
Optimal control problem 4, 9
Orthogonal 148

Pendulum 64, 74–77, 100
Piecewise constant 7
Pontryagin (*see* Maximum principle)
Principle of optimality 118

Quadratic cost 97

Rank 149
RC($\mathbf{x}$) 7
Reachable cone 7
Reachable set 7
Real canonical form 35, 148
Resource harvesting 16
Response 5
Response formula 28, 79
Rocket car 3, 10–15, 36, 64, 74, 98, 109, 111–118, 123, 125

Separate 151
Singular control 132
Smooth control 7
Spring, nonlinear 43
Stability 157
State 1
Stochastic control 22
Subspace 151
Successful control 4
Sufficient conditions 82
Supporting hyperplane 151
Symbols, list of ix
Symmetric 150
Synthesis 6, 9
System 1

$\mathcal{T}(t)$ 5
Target 2, 5
Time lags 22
Time-optimal control 57 *et seq.*
Translate 151
Transpose 148
Transversality 6, 125

$\mathcal{U}_\epsilon$ 7, 44
$\mathcal{U}_{BB}$ 7, 44
$\mathcal{U}_{BBPC}$ 7, 29, 44
$\mathcal{U}_{PC}$ 7, 44
Uniqueness of response 68

Variation of parameters 156

Water storage 1
Weak compactness 154
Weak convergence 154

Undergraduate Texts in Mathematics

(continued)

Lang: Calculus of Several Variables. Third edition.
Lang: Introduction to Linear Algebra. Second edition.
Lang: Linear Algebra. Third edition.
Lang: Undergraduate Algebra. Second edition.
Lang: Undergraduate Analysis.
Lax/Burstein/Lax: Calculus with Applications and Computing. Volume 1.
LeCuyer: College Mathematics with APL.
Lidl/Pilz: Applied Abstract Algebra.
Macki-Strauss: Introduction to Optimal Control Theory.
Malitz: Introduction to Mathematical Logic.
Marsden/Weinstein: Calculus I, II, III. Second edition.
Martin: The Foundations of Geometry and the Non-Euclidean Plane.
Martin: Transformation Geometry: An Introduction to Symmetry.
Millman/Parker: Geometry: A Metric Approach with Models. Second edition.
Owen: A First Course in the Mathematical Foundations of Thermodynamics.
Palka: An Introduction to Complex Function Theory.
Pedrick: A First Course in Analysis.
Peressini/Sullivan/Uhl: The Mathematics of Nonlinear Programming.
Prenowitz/Jantosciak: Join Geometries.
Priestley: Calculus: An Historical Approach.
Protter/Morrey: A First Course in Real Analysis. Second edition.
Protter/Morrey: Intermediate Calculus. Second edition.
Ross: Elementary Analysis: The Theory of Calculus.
Samuel: Projective Geometry.
Readings in Mathematics.
Scharlau/Opolka: From Fermat to Minkowski.
Sigler: Algebra.
Silverman/Tate: Rational Points on Elliptic Curves.
Simmonds: A Brief on Tensor Analysis. Second edition.
Singer/Thorpe: Lecture Notes on Elementary Topology and Geometry.
Smith: Linear Algebra. Second edition.
Smith: Primer of Modern Analysis. Second edition.
Stanton/White: Constructive Combinatorics.
Stillwell: Elements of Algebra: Geometry, Numbers, Equations.
Stillwell: Mathematics and Its History.
Strayer: Linear Programming and Its Applications.
Thorpe: Elementary Topics in Differential Geometry.
Troutman: Variational Calculus with Elementary Convexity.
Valenza: Linear Algebra: An Introduction to Abstract Mathematics.
Whyburn/Duda: Dynamic Topology.
Wilson: Much Ado About Calculus.

9 780387 906249